Practical Electricity

J. F. McPartland

W. J. Novak

Associate Editors, *Electrical Construction and Maintenance*

McGRAW-HILL BOOK COMPANY

New York San Francisco Toronto London

PRACTICAL ELECTRICITY

 Library of Congress Catalog Card Number 64–25173

ISBN 07-045694-1

11121314 HDBP 7543

PREFACE

This book presents the fundamentals of electricity in a unique and appealing manner. The illustration technique was particularly chosen for its easy-to-use style. Designed to provide a broad and deep, but essentially nonmathematical, background in the nature of electricity and the operation and application of electric circuits, the presentation makes use of easy-to-follow sketches and diagrams to supplement the text. Each step of the overall development is broken down into readily digested units of text and sketches.

Starting with the physics of electric current flow, the book describes and analyzes both direct-current and alternating-current electric circuits. To assure tight correlation to modern practice, examples are presented in terms of commonly used voltages and current ratings, covering circuits typical of those found in today's electrical systems for power and light in residential, commercial, and industrial buildings. The heaviest emphasis is placed on practical single-phase and three-phase alternating-current circuits and equipment.

This book can serve well as a textbook for various formal courses in basic electricity. It also offers wide use as a home-study or reference book for those engaged in design, construction, or maintenance of electrical systems. And the volume is ideally suited to the technical information needs of sales personnel and other semitechnical personnel engaged in electrical work. Its content is aimed at both the beginner and the experienced, practicing electrical man.

J. F. McPartland
W. J. Novak

CONTENTS

Basic Electricity

THE MODERN BASIS for understanding the nature of electricity is the electron theory of matter. According to this theory, all matter in the world—anything that can be seen, felt, or otherwise determined to be physical in nature—is made up of minute charges of electricity. Any physical substance can be broken down into extremely tiny particles called "atoms," which are so small that they are not visible even under a powerful microscope. Each atom is an electrically balanced package of charged particles. The core of the atom, called the "nucleus," has a given value of positive charge depending upon the element or substance it happens to be. Revolving around the nucleus are particles called electrons, each of which has a negative charge of one unit. The number of electrons "orbiting" the nucleus in any atom equals the number of units of positive charge in the nucleus. It is these electrons which make possible the flow of electric currents.

In metallic substances, the electrons in the atomic structures are normally held in their orbits by relatively weak forces of attraction from the nuclei. As a result, if a source of electron pressure (such as a battery or generator) is applied in series with a continuous path of metal, such as copper or aluminum wire, the force exerted by the pressure is sufficient to overcome the forces holding the electrons to the nuclei in each atom and these electrons are made to move through a metal path. Thus a current is created with electrons flowing from the negative terminal of the source to the positive terminal. The relative "looseness" of electrons in any metal determines its value as a conductor. Copper and aluminum are the common conductor materials used in modern electrical systems.

In contrast to the relatively good conductivity of metallic materials, there are other materials which have very "tight" electron structures and do not permit appreciable electron flow when subjected to electron pressure.

Such materials are called dielectrics or insulators. Typical of such materials are paper, rubber, plastic, porcelain, cloth, wood and just about any nonmetallic solid. Of course, such materials vary in their insulation strength, depending upon their particular atomic makeup. Materials with high resistance to flow of electric current find use as insulation on wires and cables and in electrical apparatus.

Inasmuch as electric current consists of flow of electrons, its value is measured in terms of electrons passing a point each second. This is identical to measuring flow of water in a pipe in terms of gallons per minute passing any point in the pipe. Because there is such a large number of electrons flowing in even very small currents (billions and billions of electrons per second), the unit called the "ampere" is used and is equivalent to $6.28 \text{ v } 10^{18}$ electrons per second.

The usefulness of electric current, however, is not in flowing through circuit conductors. It must be made

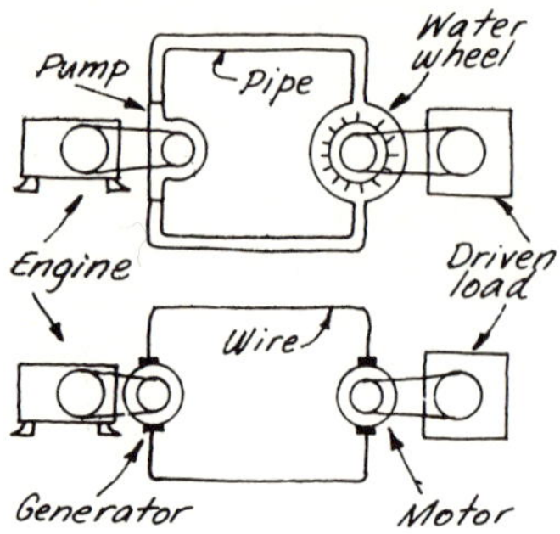

FLUID ANALOGY

Electricity is a means of transmitting energy. It is not a fluid, but components of an electrical system have counterparts in a fluid system, as shown. The flow of current in the wire (or water in the pipe) makes it possible for the driven load to make efficient use of the energy developed by the engine, located remotely from the load.

to flow through incandescent and fluorescent lamps to produce light, to flow through resistance elements to produce heat or to flow through motor windings or other coils where electromagnetic effects will be converted into mechanical work. The wires or cables in any circuit are merely the means to deliver the current to the utilization devices which will perform the work. As such, system design aims at keeping their opposition to current flow down to a minimum. The generator or battery then, which supplies the electron pressure to the circuit, has to spend only a little of its force in getting the current through the wires and can spend its major force in pushing the electrons through the load device which offers the major opposition to current flow.

Basically, any electric circuit consists of the utilization device or devices which will use the current, the conductors which will carry the current to the device and a source of pressure. This source of pressure may be a battery or a generator which will feed directly to the load or it might be the service conductors which the utility brings to a building to provide the needed pressure for all of the circuits in the buildings. Of course, these service conductors actually connect back to the utility company's generator in a remote location. In the case of a typical branch circuit for lighting or a motor, the busbars in the panelboard can be thought of as the source of pressure. In fact, any set of terminals across which a voltage exists can be considered the pressure source.

ELECTRON THEORY

When electrical pressure is exerted on a "conductor" by a generating device electrons are moved readily from atom to atom, creating an electric current.

Electrons in an "insulator" are held tightly by their atoms. Electrical pressure does not cause them to move, hence no current flow results.

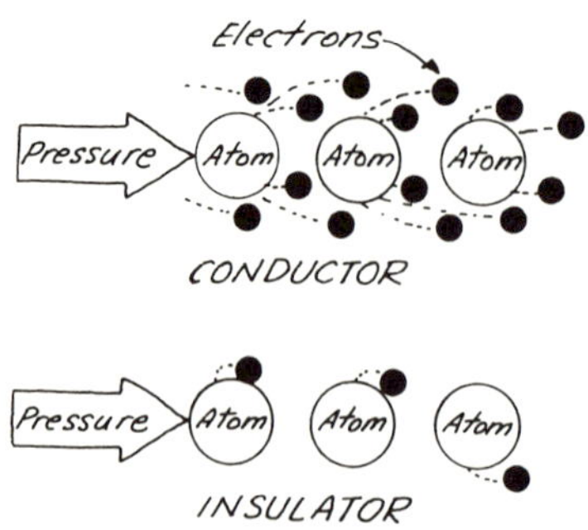

Electric pressure is expressed in terms of volts. The more volts at which a source is rated, the higher will be the pressure it exerts on the electrons in a circuit. As stated above, this pressure is used to overcome the forces holding atoms in electrical balance. The amount of force to be overcome in any part of a circuit depends upon the atomic tightness of the material itself, the cross-section area, and the length of the section concerned. This total force can be expressed as resistance or impedance to current flow. In a typical circuit, the copper or aluminum conductors have very loose atomic structures and sufficient cross-section so that the total resistance for the length of the conductor is relatively low. In contrast, the filament of a lamp being supplied by the conductors is made of a length of fine tungsten wire of very small cross-section, with the comparatively tight atomic structure of tungsten. As a result, this filament has a high resistance compared to the circuit conductors—even though length of the filament is short. Resistance or impedance to current low is commonly expressed in terms of "ohms".

From the foregoing, it can be seen that in any circuit it takes volts to push current through the resistance of conductors and load devices. In our typical electrical system, we provide 120 volts to force enough current through an incandescent lamp to produce light. Or we provide 480 volts to push enough current through a motor winding to get a certain horsepower of mechanical output.

In the case of the lamp, the current flow through the lamp filament produces tremendous molecular friction resulting in high heat. The heat causes the filament to glow "white hot," giving off visible light. The current through the motor produces magnetic effects which result in rotation of a shaft.

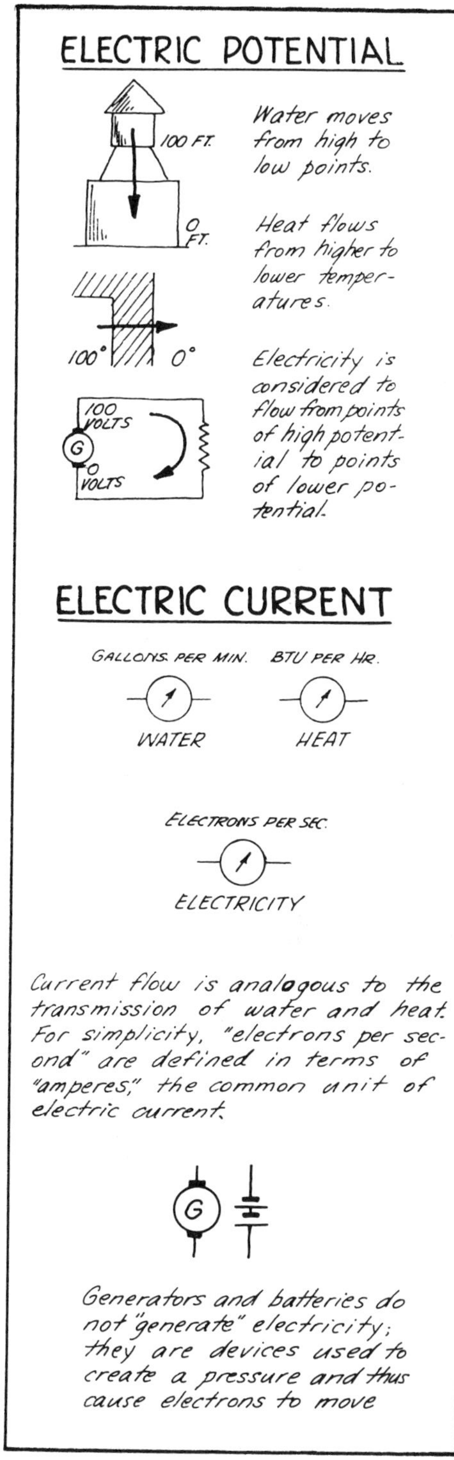

Circuits

THE FUNDAMENTAL requirement for electrical utilization is the electric circuit.

As stated previously, a circuit consists of a source of voltage (or electron pressure)—such as a battery, a generator or any terminals across which a voltage exists—one or more "load" devices which will use the current provided by the source, and a closed conductive path—such as insulated wires—from the negative terminal of the source, through the load and back to the positive terminal of the source. This arrangement is essential to all types of useful circuits:

1. The typical lighting branch circuit gets voltage pressure from the busbars in the panel; wires are run in cable or conduit to outlets where load devices—fluorescent and incandescent luminaires — are connected to the source; and current flows when the circuit is closed by a wall switch.

2. A motor circuit may get voltage from power panel busbars or from spliced connections to feeder conductors; wires complete the path to the motor; and the motor is the load.

3. A feeder circuit is any circuit in an electrical system which feeds a number of branch circuits. A typical example would be the conductors in, say, an office building which run from the main switchboard in the basement up to a panelboard on one of the floors. In such a case, the terminals on the switchboard constitute the voltage source, the conductors to the panelboard busbars make up the closed circuit path and the load is considered to be the sum of the loads of the branch circuits supplied from the panelboard.

4. A supply circuit or service entrance circuit to a building has the utility line terminals as the voltage

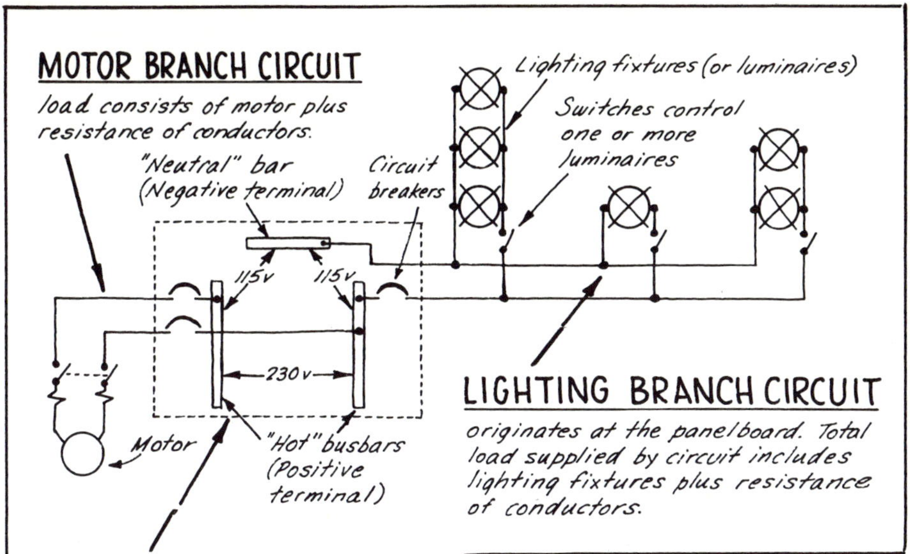

source, the conductors connecting to the building service equipment as the closed circuit and all of the load devices in the building—lights, appliances, motors, etc.—as a single effective load at the main switchboard incoming terminals.

Any of the above types of circuits might be direct-current circuits or alternating current circuits depending upon the operating character of the voltage source, whether it is ac or dc. Such circuits might operate at any of the voltage levels in use today, ranging from below 50 volts for various signal and control circuits, through the common 120-, 208-, 240-, 277-, 480- and 550-volt circuits used for lighting and power loads, to the higher voltages—2,300; 4,160; 4,800; 13,000 and up—used for distribution of large amounts of power. It should be noted that all of these circuits are normally operating, useful circuits. There is a condition which can occur in any circuit in which failure of insulation in equipment or in conductors provides a closed path of current flow that is not through the load. Such a condition is called a "short circuit."

In any electrical circuit, the flow of current depends upon the character of the voltage source and the nature of the conductive path through the entire circuit. As a result, there are different types of current flow.

Direct Current is current which flows through the circuit in the same direction at all times and with almost constant strength. A direct current flow in which the strength is always the same, such as from a battery, is often called a "continuous" current. Direct current in which the strength of flow rises and falls during the time of flow is called "pulsating" direct current. This latter current is produced by a dc generator, but if the amount of the pulsation is small compared to the overall value of the current, the current is effectively direct current.

Alternating Current is the type of circuit flow produced by a voltage source which alternates the direction of its electron "push." Say a source has terminals marked 1 and 2. Then—on a fixed time schedule, say, 120 times per second — the source will exert pressure first from 1 toward 2, then from 2 toward 1. Naturally, current in the circuit will respond to this alternating pressure and will itself alternate direction of flow in time with the alternations of pressure. Electrons will quickly flow one way, stop, reverse direction, stop, reverse direction, etc. During the period between each reversal of direction, the current starts from a value of zero amperes, rises to a maximum value of so many amperes and then decreases to zero. This alternation is the basis of our modern 60 cycles of current alternation per second. Each cycle consists of flow from zero-to-maximum-to-zero in one direction and zero-to-maximum-to-zero in the opposite direction. There are 120 half cycles and 120 reversals of flow direction per second.

Although the characteristics of alternating-current circuits differ greatly from those of direct-current circuits (and will be discussed separately later), there are a number of fundamental rules common to both types of circuits. One of the most important of these is Ohm's Law and applies to ac and dc circuits.

George Ohm, a teacher of math and physics in Germany, set forth a law in 1827 covering the behavior of electrical circuits. According to his law, a fixed relationship exists among current, voltage and resistance in a closed circuit. The law may be stated in three ways:

1. The current flowing through a circuit is equal to the voltage applied to

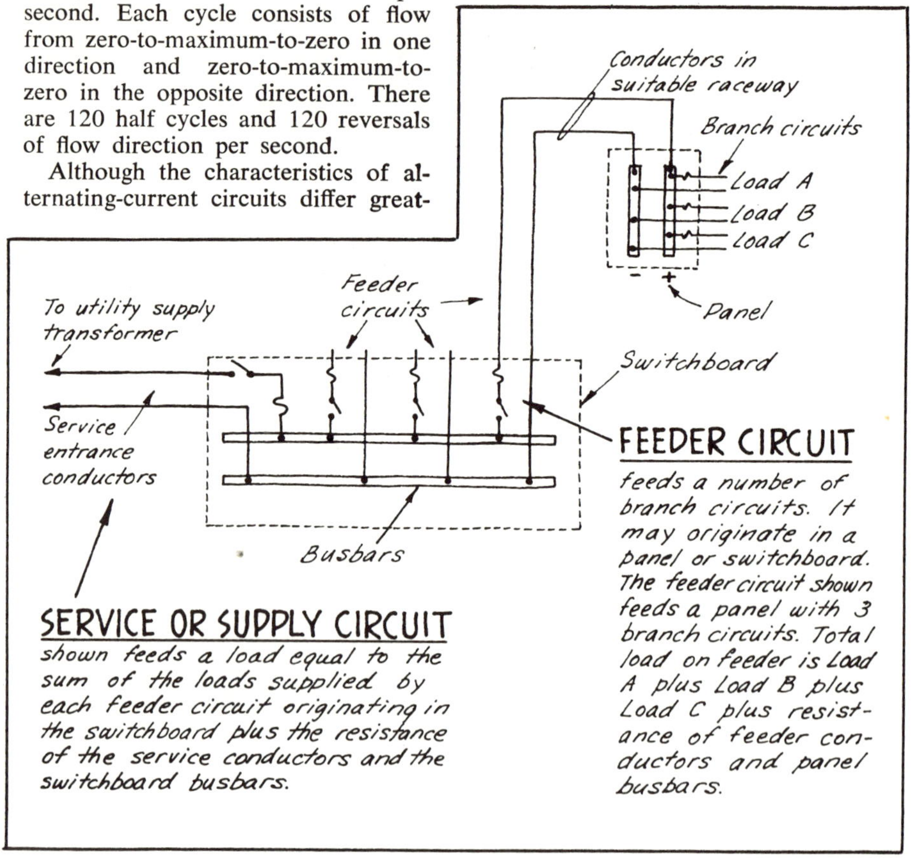

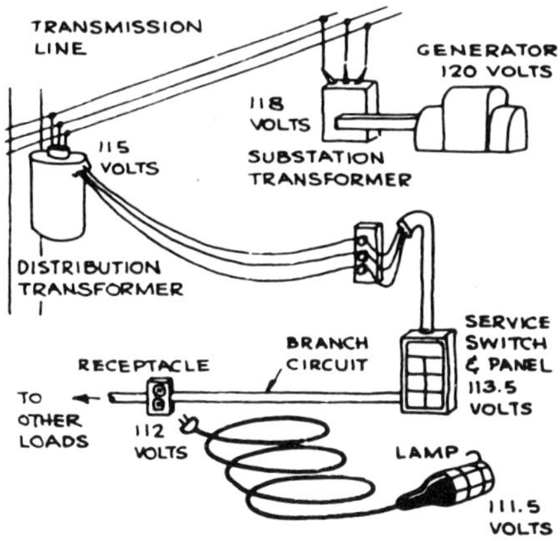

VOLTAGE DROP *in any part of an electrical circuit represents the voltage used to force the current through that part of the circuit. In the sketch at left, there is a voltage drop of 2 volts between the generator and substation transformer, 3 volts between the substation and distribution transformer, etc. Thus there is a potential of only 111.5 volts across the lamp when it is plugged into the receptacle. The voltages shown will vary, of course, with the actual current flowing in the conductors.*

Just what comprises the "internal" and "external" circuits is a matter of opinion. Basically, everything outside the generator is external to the generating source. However, from the standpoint of the occupancy served, only that portion beyond the service switch may be considered the external circuit. As far as the extension cord and lamp are concerned, everything up to the receptacle is "internal." The exact definition adopted is unimportant.

the circuit divided by the total resistance in the circuit.

$$I = \frac{E}{R}, \text{ where}$$

I=current in amperes
E=voltage applied, and
R=resistance in ohms.
If 120 volts is applied to a circuit containing a total resistance of 12 ohms, a current of 10-amps will flow.

2. The voltage applied to a circuit is equal to the current flowing through the circuit times the resistance of the circuit.

$$E = IR$$

If a current of 30 amps is flowing through a circuit with a total resistance of 16 ohms, the voltage of the circuit is 480 volts.

3. The resistance of a circuit is equal to the voltage applied to the circuit divided by the current flowing through the circuit.

$$R = \frac{E}{I}$$

If a circuit has 240 volts applied to it and the current flow is 8 amps, the total resistance is 30 ohms.

IN THE three forms of Ohm's Law— $I=\frac{E}{R}$, $E=IR$ and $R=\frac{E}{I}$ the voltage is applied to the circuit which is made up of the resistance of the circuit conductors and the resistance of the load device.

As a result, current flowing through the resistance of the conductors produces a certain voltage drop in accordance with the formula E=IR, in which the E equals the drop in voltage, the I equals the current flowing and the R equals the resistance of the cir-

SERIES CIRCUIT

Consider only that portion of the system beyond the receptacle shown. In this simple circuit, there are 3 resistances in series:

Conductor to lamp:	*0.25 ohm*
Lamp:	*121.00 ohms*
Conductor from lamp:	*0.25 ohm*
Total resistance:	*121.50 ohms*

To find the current flowing in the circuit, apply Ohm's law:

$$I = \frac{E}{R} = \frac{112 \text{ volts}}{121.5 \text{ ohms}} = 0.922 \text{ ampere.}$$

Part of the 112 volts is used up to force the current through each of the three resistances. The voltage drop across each part may be found by applying Ohm's law to each resistance in turn:

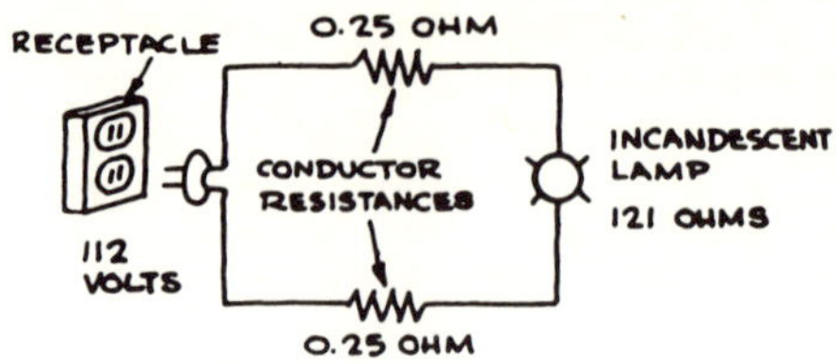

Conductor to lamp:

$E = IR = 0.922 \text{ amp} \times 0.25 \text{ ohm}$
$= 1/4 \text{ volt (approx.)}$

Lamp:

$E = IR = 0.922 \text{ amp} \times 121 \text{ ohms}$
$= 111\frac{1}{2} \text{ volts (approx.)}$

Conductor from lamp:

$E = IR = 0.922 \text{ amp} \times 0.25 \text{ ohm}$
$= 1/4 \text{ volt (approx.)}$

Thus, in a series circuit,

- THE SAME CURRENT FLOWS IN ALL PARTS.
- THE VOLTAGES ACROSS EACH PART ADD UP TO THE TOTAL IMPRESSED VOLTAGE.

cuit conductors. The voltage dropped in the conductors is then subtracted from the voltage applied to the circuit to obtain the voltage which is actually impressed across the terminals of the lamp, motor or other load device. Such consideration of voltage drop is an important phase of any actual circuit calculation and will be considered fully later.

Because circuits commonly supply a number of lamps or a number of motors or other power devices, it is important to be able to calculate the values of current, voltage and resistance for such operating conditions. In the common arrangement, load devices are rated for the circuit voltage — such as 120 volts — and are connected across the conductors in parallel. Each load device is then supplied with its rated voltage and draws current in proportion to the value of its resistance. The circuit current is then equal to the sum of the currents drawn by the load devices connected in parallel. Or another way to calculate the circuit current is to reduce the resistances of the parallel load devices to a single resistance value equivalent to the several.

Calculation of circuit conditions for supplying a number of independent load devices is based upon two rules of load connections:

1. If two or more resistance load devices are connected in series, i.e., the load resistances are connected in series with the resistance of the circuit conductors, the effective load resistance is equal to the sum of the individual load resistances. And the total circuit resistance is the sum of the effective load resistance and the resistance of the circuit conductors. A circuit in which resistances are connected in a single path of current flow, with current flowing from one resistance to the next as it makes its way from the positive terminal to the negative terminal of the voltage source, is known as a series circuit.

2. If two or more resistances are connected so that each resistance has the full circuit voltage impressed across it, the resistances are said to be connected in parallel. In such an arrangement, the effective resistance of all of the resistances considered as a single resistance is equal to some value less than the smallest of the resistances connected in parallel.

Series Circuits

As far as the connection of load devices is concerned, series circuits are not generally used for modern lighting and power load connection. The theory of series circuit operation, however, is essential to a full understanding of what takes place in every branch circuit, feeder and service supply. As mentioned previously, every circuit finds the circuit wires themselves as resistances in series with the single or combination load resistance.

In any strictly series circuit, the voltage at the source of the circuit is impressed across the sum of the resistances. As a result, the current that flows in the circuit can be determined from Ohm's Law as follows:

$$I = \frac{E}{R_1 + R_2 + R_3 + R_4 + R_5 + R_6 + \text{etc.}},$$

in which

I=current flowing through all of the resistances

E = the circuit voltage

R_1, R_2, R_3, R_4, etc. = the individual resistances making up the circuit.

Then, applying one of the other forms of Ohm's Law to each resistance —i.e., E=IR—we will get the amount of the total circuit voltage which is dropped across each resistance, the amount of electron pressure needed to produce the current flow through each resistance. And the sum of these voltage drops around the circuit—$E_1 = IR_1$, $E_2 = IR_2$, $E_3 = IR_3$, etc. —is equal to the voltage impressed at the source of the circuit.

It should be noted that the current through a series circuit has the one value determined by the impressed voltage and the sum of the resistances. It is the same at any point in the circuit. Voltages across individual resistances, on the other hand, vary with the value of resistance and the amount of current flowing.

Of course, it must always be remembered that the voltage source itself—whether it is a generator or a battery, or the busbars in a switchboard or panelboard—has a definite amount of resistance, known as "internal" resistance, which it adds to the series resistances of the external circuit and which plays a part in limiting current flow. This latter characteristic requires that a careful distinction be made between the parts of any circuit, as follows:

A. That part of the circuit which is connected to the terminals of the source—to the generator terminals or to the bus lugs in a panelboard—is generally referred to as the "external" circuit. The resistances of the external circuit include the load resistances and the resistances of connecting wires from the source to the load and between individual load resistances in a series circuit.

B. That part of any circuit which is within the generator or battery or that part which is between a set of terminals used as a voltage source and the actual generator or other source providing the voltage at the terminals can be thought of as the internal circuit. This internal circuit contains the electromotive force, or electron pressure, and the resistance of its own current path. This internal resistance reacts with current flowing through it to produce a certain voltage drop which substracts from the voltage value of the electromotive force. As a result, if the voltage across the terminals of a generator or other source is measured when the voltage source is not supplying current to a closed external circuit, the value of voltage will be exactly equal to that generated.

If, however, the voltage of the same source is measured when the source is supplying current to an external circuit, the voltage will be found to be somewhat lower than the open-circuit value. The difference in the two measured values will be equal to the product of the amperes of current flowing times the value of the internal resistance of the source—according to Ohm's Law that E=IR. It should be noted that voltage sources produce electrical pressure, but current flow through resistance always produces a drop in pressure. That is, it uses pressure and must therefore be subtracted from the source value.

Any voltage drop, then, in the internal resistance of a source must be subtracted from the value of generated electromotive force to determine the actual value of voltage available at the terminals of the source to produce current flow through the external circuit.

APPLICATION of series-circuit calculations takes many forms in modern electrical practice. Although utilization devices—lighting fixtures, motors, appliances—are generally connected in parallel on branch circuits, street lighting circuits are frequently connected with the luminaires or ballasts in series. Series calculations are also standard parts of engineering design of feeders and of many motors and appliances. **Some typical examples of calculations are as follows:**

1. A generator with an internal resistance of 0.5 ohm produces an electromotive force of 120 volts and is connected by two wires, each with a resistance of 1 ohm, to a load device with a resistance of 57.5 ohms. What is the circuit current? What is the terminal voltage of the generator? What is the voltage drop in the circuit conductors? What is the voltage across the load?

a. The current flowing through the series circuit is equal to the emf divided by the total resistance.

$$I = \frac{120 \text{ volts}}{0.5 + 1 + 1 + 57.5} = \frac{120 \text{ volts}}{60 \text{ ohms}} = 2 \text{ amps}$$

b. Generator terminal voltage is equal to the generated emf minus the voltage drop produced by the circuit current and the internal resistance of the generator.

$$\text{Terminal voltage} = 120 - (2 \text{ amps} \times 0.5 \text{ ohms}) = 119 \text{ volts}$$

c. Voltage drop in the circuit conductors is equal to circuit current times the total resistance of the conductors.

$$\text{Voltage drop} = 2\text{amps} \times 2 \text{ ohms} = 4 \text{ volts}$$

d. The voltage across the load is equal to the generator terminal voltage minus the voltage drop in the circuit conductors.

$$\text{Load voltage} = 119 \text{ volts} - 4 \text{ volts} = 115 \text{ volts}$$

2. Four lamps, each rated at the same resistance value, are series connected across a terminal voltage of 600 volts, producing a current flow of 10 amps. What is the circuit resistance? What is the resistance of each lamp? What is the voltage across each lamp?

a. Circuit resistance is equal to the terminal voltage divided by the circuit current.

$$R = \frac{600 \text{ volts}}{10 \text{ amps}} = 60 \text{ ohms}$$

b. Since the four lamps are connected

PARALLEL CIRCUIT *consists of coffee maker, toaster, and frying pan plugged into kitchen appliance circuit. What currents will flow in the portions of the circuit indicated by the arrows in the drawing? Neglect the voltage drop in the conductors.*

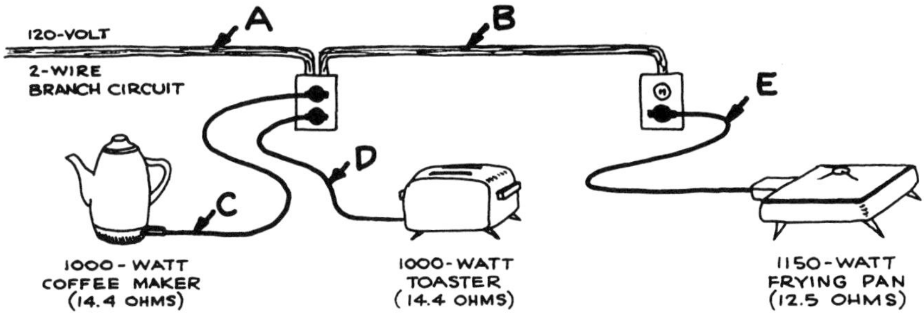

THE FIRST STEP *is to draw a simple schematic diagram to represent the circuit. Since we are neglecting the conductor voltage drop, there is a 120-volt potential across each appliance.*

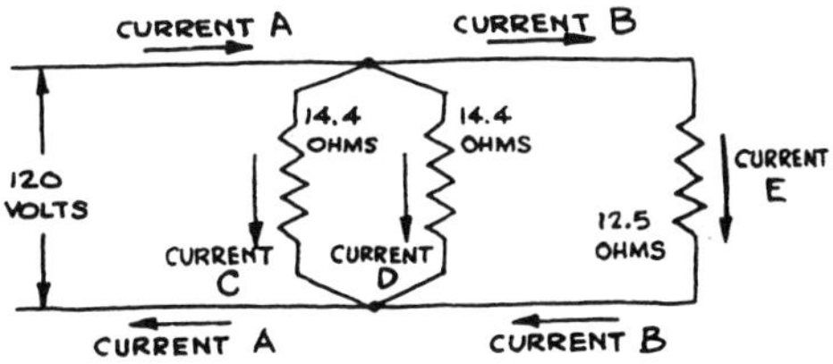

We may apply Ohm's Law in turn to each appliance:

$$\text{CURRENT} = \frac{\text{VOLTS}}{\text{OHMS}}$$

$$C = \frac{120 \text{ VOLTS}}{14.4 \text{ OHMS}} = \boxed{8.33 \text{ AMPERES}}$$

$$D = C = \boxed{8.33 \text{ AMPERES}}$$

$$E = \frac{120 \text{ VOLTS}}{12.5 \text{ OHMS}} = \boxed{9.6 \text{ AMPERES}}$$

$$B = E = \boxed{9.6 \text{ AMPERES}}$$

Current A, the total branch circuit current, is equal to the sum of the currents in each of the parallel branches:

$$A = C + D + E$$

$$A = 8.33 + 8.33 + 9.6$$

$$A = \boxed{26.26 \text{ AMPERES}}$$

We may also find current A by determining R_{eff}, the effective (or equivalent) resistance:

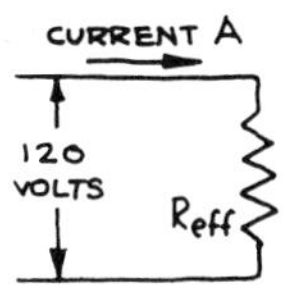

$$\frac{1}{R_{eff}} = \frac{1}{R_1} + \frac{1}{R_2} + \frac{1}{R_3}$$

$$= \frac{1}{14.4} + \frac{1}{14.4} + \frac{1}{12.5}$$

$$= .0694 + .0694 + .08$$

$$= .2188$$

$$R_{eff} = \frac{1}{.2188} = 4.57 \text{ OHMS}$$

$$A = \frac{120 \text{ VOLTS}}{4.57 \text{ OHMS}} = \boxed{26.26 \text{ AMPERES}}$$

Obviously, breakers or fuses will open the normal 20-ampere circuit with such a load. This emphasizes the desirability of having two 20-amp kitchen appliance circuits.

in series and all have the same resistance, the resistance of each is equal to the total resistance divided by the number of lamps.

$$R = \frac{60 \text{ ohms}}{4} = 15 \text{ ohms}$$

c. The voltage across each lamp is equal to the resistance of the lamp times the current flowing.

$$\text{Lamp voltage} = 15 \times 10 = 150 \text{ volts.}$$

Parallel Circuits

As stated previously, the most common connection for light and power utilization devices connected to branch circuits in modern electrical systems is the parallel connection. In such a circuit connection, all of the load resistances are connected between the conductors which connect the load to the source of emf. As a result, the voltage across any one load resistance is essentially the same as that across any other load resistance.

The voltage across all of the resistances is exactly the same if all of the resistances are connected to the same points on the circuit conductors tieing back to the emf source. Such connection, however, is not generally achieved in typical circuit hookups. In the usual arrangement some of the load resistances are further away from the switch or panelboard terminals than are others of the load resistances. There is, then, a very slight difference in voltage from one resistance to any other due to voltage drop in the conductors of the circuit. The basic theoretical point, however, is that voltage across all of the resistances is the same in parallel hookup. This can be thought of as comparing to the fact that current through all of the resistances is the same in series hookup.

With the same voltage across all of the load resistances, the total current drawn by the circuit is equal to the sum of the individual currents drawn by the individual resistances. The current divides among the parallel paths presented by the resistances, with the current through any path conforming to I=E/R. The load with the lowest resistance value will draw the highest current and that with the highest resistance value will draw the lowest current. Intermediate values of resistance will draw intermediate values of current. If, for instance, three load devices are connected in parallel on a 120-volt circuit, with the resistance values being 2, 4 and 6 ohms, the 2-ohm device will draw 120/2 or 60 amps, the 4-ohm device will draw 120/4 or 30 amps and the 6-ohm device will draw 120/6 or 20 amps. The total current drawn by the circuit conductors which supply the three devices is equal to 60+30+20 or 110 amps.

Another way of evaluating a parallel circuit involves reducing the group of parallel-connected resistances to a single equivalent resistance which would draw current equal to the sum of currents flowing through the individual resistances of the actual circuit under study. The effective resistance of a group of resistances connected in parallel will always have an ohmic value less than that of the smallest resistance of the group. For any parallel group of resistances, the actual value of effective resistance can be readily found from the following formula:

$$\frac{1}{R_{EFF.}} = \frac{1}{R_1} + \frac{1}{R_2} + \frac{1}{R_3} \cdots \text{etc.,}$$

where $R_{EFF.}$ = the effective resistance of the group, and R_1, R_2, R_3, etc. = the individual resistances

Example: Three resistances—10 ohms, 20 ohms and 40 ohms are connected in parallel. What is the equivalent single resistance value of this group?

$$\frac{1}{R_{EFF.}} = \frac{1}{10} + \frac{1}{20} + \frac{1}{40} = \frac{4+2+1}{40} = \frac{7}{40}$$

$$R_{EFF.} = \frac{40}{7} = 5.71 \text{ ohms}$$

Example: Four resistances — each rated 20 ohms—are connected in parallel. What is the effective resistance of this group?

$$\frac{1}{R_{\text{EFF.}}} = \frac{1}{20} + \frac{1}{20} + \frac{1}{20} + \frac{1}{20} = \frac{4}{20}$$

$$R_{\text{EFF.}} = \frac{20}{4} = 5 \text{ ohms}$$

From the latter example, it can be seen that when a number of parallel resistances have the same ohmic value, the effective resistance of the group is equal to the ohmic value of one of the resistances divided by the number of resistances connected in parallel.

In the first example above, if the group of resistances were connected across a circuit voltage of 120 volts, the current flowing through the circuit conductors would be E/R or 120/5.71, which is 21 amps. The current flowing through each of the individual resistances would be: 120/10 ohms or 12 amps; 120/20 ohms or 6 amps; and 120/40 ohms or 3 amps. The total of the three currents is, of course, also 21 amps.

Circuit Studies

ALTHOUGH circuits are divided into two basic categories—series and parallel—based on operating characteristics, analysis and application of modern circuits for power, light, heat, signals and communications almost always involves both series and parallel considerations.

Take the simple case of a kitchen appliance circuit which supplies, say, a coffee maker, a toaster and an electric frying pan. The appliances are connected by cord and plug to receptacle outlets. In each appliance, the heating elements may be connected in parallel. But this parallel load is actually in series with the resistance of the conductors in the power cord. Thus the circuit to each appliance from the wall outlet is a combination series-parallel circuit.

Then consider the branch circuit from the panelboard to the wall outlets. The three appliances—each a series-parallel circuit, as we have seen—are connected in parallel across the branch circuit conductors. But again, we have the resistance of the branch circuit conductors connected in series with the parallel load of three appliances. And if the electrical system is traced further back toward the generator or other source of supply, it will be found that there is a continuing succession of series and parallel elements associated with each other.

A diagram of almost any electric supply system shows a tree-like development of series and parallel circuits, with the generator as the trunk of the radial branches and subbranches. The generator produces the voltage which causes current flow through the entire system. It also supplies the total amount of current required by all of the load devices—lamps, motors, etc.—connected at the extremities of the system.

In flowing through the system, this total current is subdivided among various parallel circuits. And as any current flows to any part of the system, there is a voltage drop in accordance with the series formula $E=IR$. The voltage is, therefore, maximum at the generator and is successively reduced from point to point as current flows to the loads.

Calculation of voltage, current and resistance values in series-parallel circuits is a simple matter of correlating separate series and parallel calculation in accordance with the basic relationships for each type of circuit. With given circuit voltage and given resistance values, the first step in calculating currents is to determine the effective resistance of each parallel group of resistances and then treat the circuit as a purely series circuit.

Take, for instance, the simple case of a 4-ohm resistance connected in series with two parallel resistances of 2 ohms each, with the combination circuit supplied from a 120-volt source:

Considering the two 2-ohm resistances in parallel, we know that **when a number of parallel resistances have the same ohmic value, the effective resistance of the group is equal to the ohmic value of one of the resistances divided by the number of resistances connected in parallel.** From this, the effective resistance of the two 2-ohm resistances in parallel is 1 ohm, and this is connected in series with the 4-ohm resistance across the 120 volts.

Dividing the 120 by the total of 5 ohms resistance (1 ohm plus 4 ohms), we get a current of 24 amps flowing from the source and through the series-parallel combination of resistances. This 24 amps flows through the 4-ohm resistance and through the parallel group of 2-ohm resistances. Since the two parallel resistances have the same value, the current will divide equally in flowing through the two paths. A current of 12 amps will flow through each of the parallel resistances. Note that the current is 24 amps up to the parallel group, becomes 12 amp through each resistance and then becomes a single current of 24 amps where it flows from the two parallel resistances to return to the voltage source.

Considering the above example further, the current flow of 24 amps

DISTRIBUTION SYSTEMS

are tree-like combinations of series-parallel circuits.

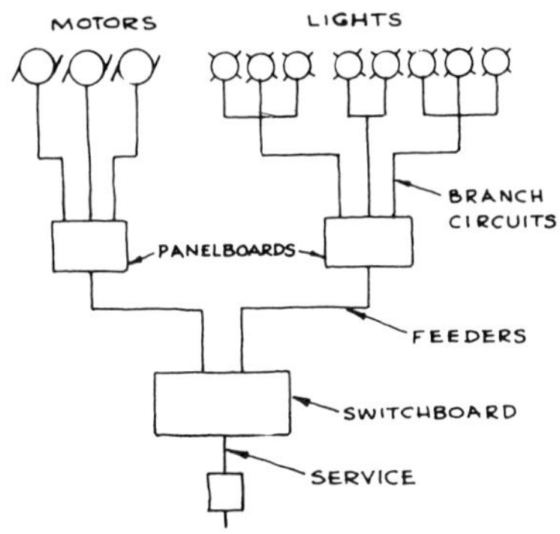

SIDEWALK HEATING CABLE

is a practical example of a series-parallel circuit:

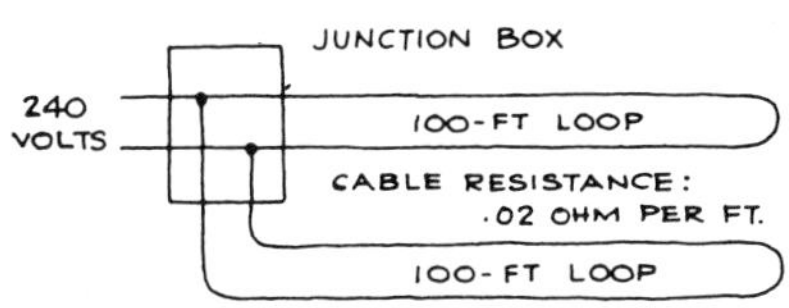

Each loop may be considered to be 100 .02-ohm resistances in series; two loops fed from a single junction box are in parallel.

TOTAL RESISTANCE OF EACH LOOP:

$100 \times .02 = 2$ OHMS

EQUIVALENT RESISTANCE OF TWO LOOPS IN PARALLEL:

$$\frac{2 \text{ OHMS}}{2} = 1 \text{ OHM}$$

TOTAL CURRENT:

$$\frac{240 \text{ VOLTS}}{1 \text{ OHM}} = 240 \text{ AMPS}$$

CURRENT IN EACH LOOP:

$$\frac{240 \text{ AMPS}}{2} = 120 \text{ AMPS}$$

VOLTAGE DROP IN EACH LOOP:

$$\frac{240 \text{ VOLTS}}{100 \text{ FEET}} = 2.4 \text{ VOLTS PER FT}$$

PROBLEM:

Find total current taken by series-parallel circuit shown; current through each resistance; voltage across each resistance.

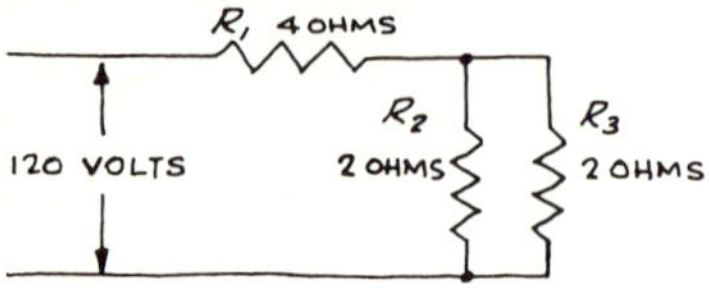

SOLUTION:

EQUIVALENT RESISTANCE OF TWO 2-OHM RESISTANCES:

$$\frac{2 \text{ OHMS}}{2} = 1 \text{ OHM}$$

EQUIVALENT CIRCUIT:

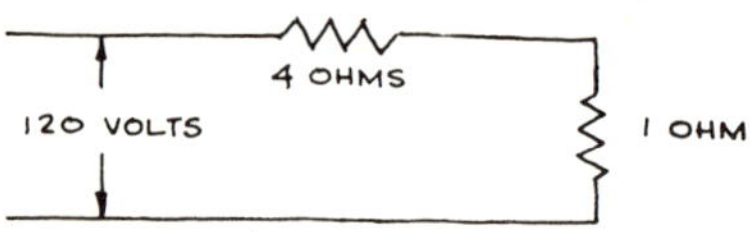

TOTAL RESISTANCE OF RESULTING SERIES CIRCUIT:

4 + 1 = 5 OHMS

TOTAL CURRENT:

$$\frac{120 \text{ VOLTS}}{5 \text{ OHMS}} = 24 \text{ AMPS}$$

TOTAL CURRENT DIVIDES THROUGH R_2 AND R_3:

$$\frac{24}{2} = 12 \text{ AMPS EACH}$$

VOLTAGE ACROSS R_1:

4 OHMS x 24 AMPS = 96 VOLTS

VOLTAGE ACROSS R_2 AND R_3:

2 OHMS x 12 AMPS = 24 VOLTS

THESE TWO VOLTAGES ARE IN SERIES:

TOTAL VOLTAGE DROP =

96 + 24 = 120 VOLTS

= IMPRESSED VOLTAGE

through the 4-ohm resistance produces a voltage drop of 4x24 (from E=IR) or 96 volts. The current flow of 24 amps through the 1-ohm effective resistance of the parallel 2-ohm resistances produces a drop of 1x24 or 24 volts, which, added to the previous 96 volts, accounts for the 120-volt pressure of the source.

Still another evaluation which can be placed on the foregoing problem is to subtract the 96 volts (4x24) from 120 volts, indicating that 24 volts are impressed across the parallel resistances. Then dividing this 24 volts by the 2-ohm value of each resistance, we get 12 amps as the current flowing through each of the parallel resistances. This, of course, is a check on the previous calculation. In such problems, the basic formulas of Ohm's Law provide for double checking to minimize the chance of error.

Although the foregoing was a simple matter of arithmetic calculating, the most complex circuits, can be reduced to similar step-by-step analysis and calculation based on the fundamental relationships—

$$I = \frac{E}{R}, \quad E = IR, \quad R = \frac{E}{I}.$$

The procedure for solving series-parallel circuit configurations can be applied to all or any part of an electrical system. In our present-day power and light systems, the feeder and branch circuit conductors are considered to be series resistances. So also are tap conductors which supply individual load devices, such as taps to lighting fixtures. Load devices—lights, motors, etc.—are generally connected in parallel with each other. However, at any panelboard, the branch circuits or feeders are connected in parallel on the busbars. At switchboards, the outgoing circuits are connected in parallel with each other.

And the feeder to a panelboard or the service conductors to a switchboard are resistance values connected in series with the effective resistance of the parallel load circuits. For any voltage level, either the terminal voltage of the utility's service or the secondary terminal voltage of a transformer that is used to step from one voltage level to another, the same approach is used.

Power and Work

CHARACTERISTICS and application of electrical circuits have, so far, been described in terms of voltage, resistance and current. These elements of electricity are, of course, important and essential to effective use of electrical systems. They are, however, strictly speaking, only incidental to the primary purpose of any electrical circuit. The function of an electric circuit is to provide power to do work.

Electrical systems, whether a simple battery operating a bell or a vast multi - voltage - level industrial layout feeding numerous power and light loads, exist only to do work—to ring a bell; to make lamps give off light; to make motors rotate with enough power to drive conveyors, pumps, fans, hoists and other mechanical loads; to make heating elements give off heat. The accomplishment of these tasks does involve extensive consideration of voltage, resistance and current; but the ultimate considerations are always of power and work. A clear understanding of the words "power", "work" and "energy" and their relationship to each other and to voltage resistance and current is essential to effective application of circuits.

Work is said to be done when a force acts through a distance. Force is understood to be a push or pull which changes the condition of motion or changes the shape of a body and is measured in standard units called **pounds.** It takes force to move a body that is at rest, to change the velocity of a moving body, to stop or change the direction of a moving body, or to deform—stretch or compress—a body. Lifting a 200 pound weight 10 feet is work rated at 2,000 foot-pounds. Or if an object were pushed along the floor with an average force of 20 pounds for a distance of 100 feet, 2,000 foot-pounds of work would be done.

It is important to note that time is not a consideration in evaluating work. In the two cases cited in the previous paragraph, the lifting of the 200 pound weight through 10 feet might have been done in, say, 5 seconds, whereas the pushing of the object by a 20 pound force over 100 feet might have taken a full minute. The work is the same in both cases, 2,000 foot-pounds. However, in the first case, the the work is being done faster. And the time-rate of doing work—how fast or

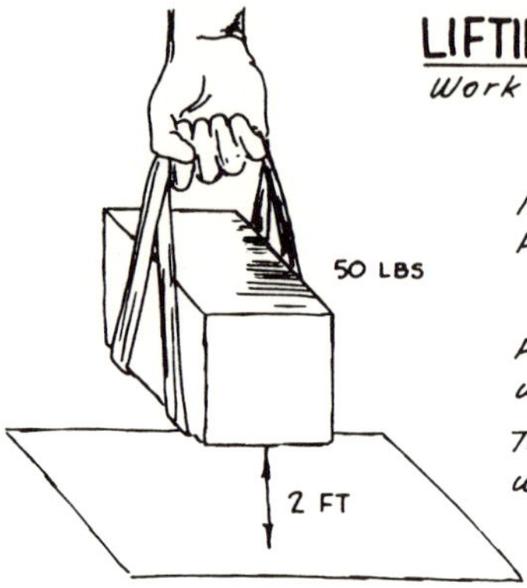

LIFTING WEIGHT *is an example of doing WORK.*
Work done in lifting 50 lbs 2 ft is

$$50 \times 2 = 100 \text{ ft-lbs.}$$

If this work is done in 5 seconds, the POWER involved is

$$\frac{100}{5} = 20 \text{ ft-lbs per second.}$$

Power is described as the rate of doing work.

Thus it would take more power to lift the weight in 2 seconds:

$$\frac{100}{2} = 50 \text{ ft-lbs per sec.}$$

how slow a given amount of work is done—is defined as "power".

Power is a measure of the rate at which work is done. A 220-pound man running up a 20-foot high flight of stairs in 10 seconds applies a lot more power than another man of the same weight who does the same amount of work in going up the same stairs in 50 seconds. Mechanical power is measured in units called "horsepower". One horsepower is equal to a work rate of 550 foot pounds per second. In the case of the two men running upstairs the first man did 220x20 or 4,400 foot-pounds of work in 10 seconds, which is 440 foot-pounds per second or 44/55 horsepower. The second man did the same 4,400 foot-pounds of work in 50 seconds, which is 88 foot-pounds per second or 88/-550 horsepower. The difference in power applied accounts for the time difference in doing the same work.

Energy is the ability to do work and is measured in the same units as work, foot-pounds. Energy is stored in a body and is expended when that body does work. In our previous discussion of work, the force applied to do work has to come from a source of energy. A source of energy is required, say, to lift a weight to a certain height. The lifting is an expenditure of energy which is work. And as a result of the lifting, the body lifted has been given potential energy equal to the work done. It can be seen that work consists in taking energy from one body and giving it to another body. Energy itself cannot be destroyed. It exists in various forms—heat, light, mechanical, electrical, chemical—and can be converted from one form to another. But the term "energy" should be differentiated from the term "work" which is the expenditure of energy.

Translating the foregoing physical concepts to conditions encountered in electrical circuits, we get the following:

Watt and kilowatt are the two terms commonly used to designate electrical power. A watt is a measure of the time-rate at which electrical energy is expended, a measure of the rate of electrical work. A kilowatt is simply 1,000 watts. When speaking of "electrical power" then, we refer to the watts or kilowatts of the load or circuit. A 1,000-watt lamp uses electrical energy 10 times as fast as a 100-watt lamp. A 20-kw generator has more power than a 10-kw generator, i.e., it provides a higher ability to do work over any period of time.

Although electrical lighting and heating devices are commonly rated in watts and circuits, systems and electrical energy sources are also rated in watts, electric motors are correlated to the mechanical loads they drive and are rated in output horsepower. One horsepower is equivalent to 746 watts or approximately ¾ of a kilowatt.

On a purely electrical basis, 1 watt is the rate of energy expenditure when a steady current of one ampere flows under a pressure of one volt. In electrical dc circuits this is expressed as a basic power equation:

$$P = E \times I$$

where, P = power in watts

E = Voltage across load

I = Current through load

For example, if there is a current flow of 0.833 amps through a lamp when 120 volts is impressed across it, the electrical power of the lamp is—

$$P = 0.833 \times 120 = 100 \text{ watts.}$$

ENERGY *is the ability to do work.*

This lamp bulb is rated 100 watts. Its power consumption is 100 watts. If it operates for 20 hrs, the energy expended is 100 x 20 = 2000 watt-hours or 2 kilowatt-hours.

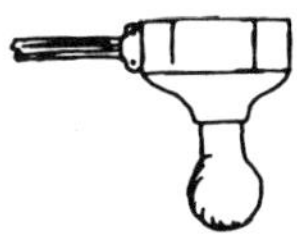

100 WATTS

EXAMPLES:

1. What is the resistance of a 60-watt, 120-volt lamp bulb?

$$\text{OHMS} = \frac{\text{VOLTS}^2}{\text{WATTS}} = \frac{120 \times 120}{60} = \frac{14400}{60} = \underline{240 \text{ OHMS}}$$

2. What current flows through this lamp?

$$\text{AMPS} = \frac{\text{WATTS}}{\text{VOLTS}} = \frac{60}{120} = \underline{0.5 \text{ AMP.}}$$

3. A 1000-watt coffee-maker operates for 3 hours each day. How much energy is expended per week?

$$\text{KWHRS} = \text{KW} \times \text{HOURS}$$
$$= 1 \text{ KW} \times 3 \times 7$$
$$= \underline{21 \text{ KWHRS}}$$

4. An electric water heater is rated 3 kw, 240 volts. What power does it require operating at 120 volts?

$$\text{OHMS} = \frac{\text{VOLTS}^2}{\text{WATTS}} = \frac{240 \times 240}{3000} = 19.2$$

Operating at 120 volts, the resistance is still 19.2 ohms:

$$\text{POWER} = \frac{\text{VOLTS}^2}{\text{OHMS}} = \frac{120 \times 120}{19.2} = \underline{750 \text{ WATTS.}}$$

Thus, only 1/4 the rated power is required to operate the heater at 1/2 rated voltage.

POWER AND ENERGY RELATIONSHIPS

$$\text{WATTS} = \text{VOLTS} \times \text{AMPS}$$

$$\text{WATTS} = \text{AMPS}^2 \times \text{OHMS}$$

$$\text{WATTS} = \frac{\text{VOLTS}^2}{\text{OHMS}}$$

1 HORSEPOWER = 550 FT-LB/SEC.
1 HORSEPOWER = 746 WATTS
1 HORSEPOWER = 0.746 KW
1 KILOWATT = 1.34 HORSEPOWER

INASMUCH as the basic equation for determining the power of electrical loads—P=E x I involves the voltage and current associated with the load, the equation can be written other ways by substituting equivalents for voltage and current from the equations of Ohm's Law:

$P = E \times I$
but, $E = I \times R$
therefore, $P = I \times R \times I = I^2 R$

If we have a load of 10 ohms resistance with a current of 4 amps flowing through it, the power of the load is

$P = 4^2 \times 10 = 160$ watts.

Or by substituting for I in the basic equation in accordance with Ohm's Law that I=E/R, we get

$$P = E \times E/R = \frac{E^2}{R}$$

Then if we had a load of 20 ohms resistance with 120 volts impressed across it, the power of the load would be

$$P = \frac{120^2}{20} = 720 \text{ watts}$$

Although the basic power equation is stated as P = E x I, to determine the power when voltage and current are known, it can be rearranged to yield any of the three quantities when the other two are known.

Example: What is the voltage across a resistance heating element rated at 1,200 watts through which 10 amps are flowing?
E = P/I = 1,200 watts/10amps = 120 volts.

Example: What is the current flowing through a load rated at 2,400 watts with 120 volts impressed across its terminals?
I = P/E = 2,400 watts/120 volts = 20 amps.

Electric Energy

As we have seen, power is the time rate of doing work. Power rating tells how fast or how slow a load will expend electrical energy or, which is the same thing, do work. Electrical work, then, is expressed in units which represent the product of electrical power and time. The basic unit of work or energy expenditure is the watthour. A watthour is simply the energy expended when power of one watt is applied for one hour. A kilowatt-hour is 1,000 watthours.

Example: If an electric light bulb is doing work at the rate of 100 watts (or, as it would more commonly be expressed, if the bulb is rated at 100 watts) and operates for 20 hours, the work done or the energy expended would be:
Energy=100 watts x 20 hours = 2,000 watthours or 2 kilowatthours.

Although the concepts of electric power and energy are applied primarily to load devices—lights, heaters, motors—to determine desired utilization, there are undersirable, though inevitable, power and energy losses in electrical systems. As we saw previously, current flow over conductors produces an unavoidable drop in voltage from the source to the load. Similarly, as long as the current flows to the load there are power losses proportional to the volts dropped in the conductors and the current flowing. And for any period of time over which power is applied, there is an expenditure of energy or an energy loss equal to the product of the power loss and the time involved. Both of these losses are inevitable in every system, but care in design is usually taken to minimize such losses. These losses not only cost money due to the charge for watthours expended in the conductors, but the electrical energy is expended by conversion to heat which can have deleterious effects on the insulation of the conductors.

Basic Meters

In practice, application of foregoing theory to circuits and equipment

METER CONNECTIONS

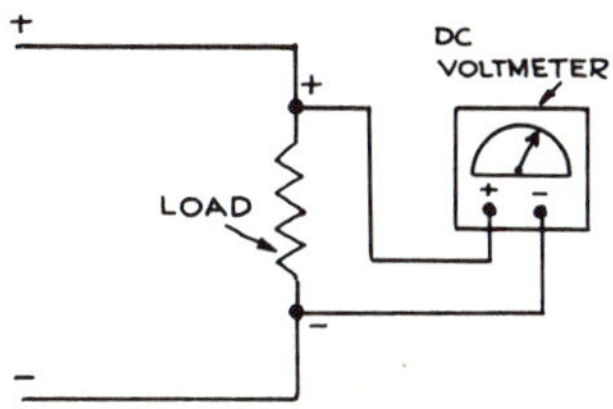

The VOLTMETER has two terminals, is connected by two conductors directly across load to measure the voltage, or potential difference, across the load. In dc circuits, as shown, the positive terminal of the meter must be connected to the positive side of the load. In ac circuits, either terminal of an ac voltmeter may be connected to either side of the load.

The AMMETER has two terminals, is connected in series with the load to measure the current flowing in the circuit. The CLAMP-ON AMMETER has hinged jaws which are slipped over conductors. No physical connection is required.

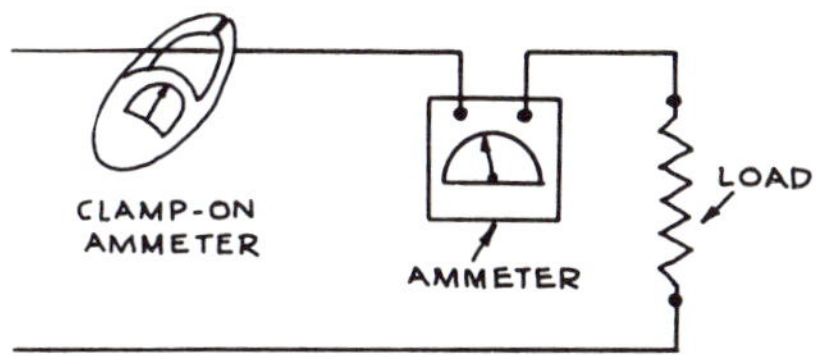

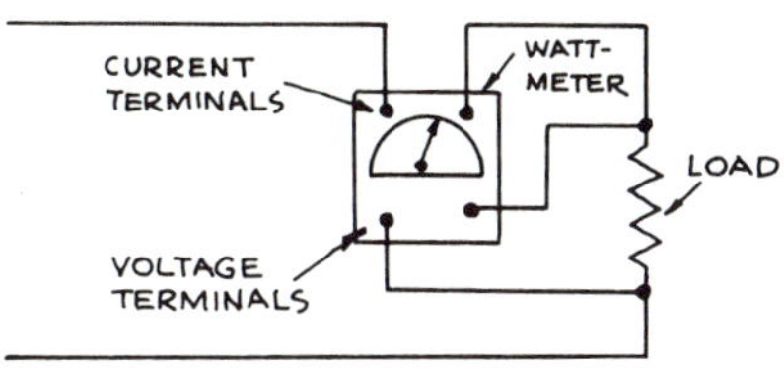

The WATTMETER combines functions of ammeter and voltmeter to give a reading directly in watts. It has two current terminals, connected in series with load, and two voltage terminals, connected across the load.

The WATTHOUR METER (or kilowatthour meter), like the wattmeter, has both voltage and current terminals. A rotating disk, or armature, operates indicating dials or a counter to show the total energy consumed. It is normally connected through a meter socket at the service entrance location.

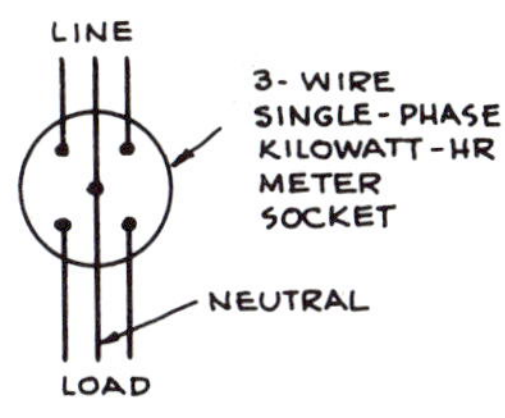

found in any electrical system depends upon some means of measuring unknown conditions and correlating the data obtained. A basic group of instruments provide for ready and accurate measurements of electrical conditions. These are:

Voltmeter—This is a device constructed and calibrated to provide directly-read indication of value of voltage when properly applied. A voltmeter must always be connected across in parallel with the load device, the circuit or circuit element across which the difference in electrical potential is to be determined. Proper connections in dc circuits call for the positive terminal of the meter to be connected to the high-potential side. Care must also be taken to use a meter with a range (as indicated on its scale) suitable for

EXAMPLES

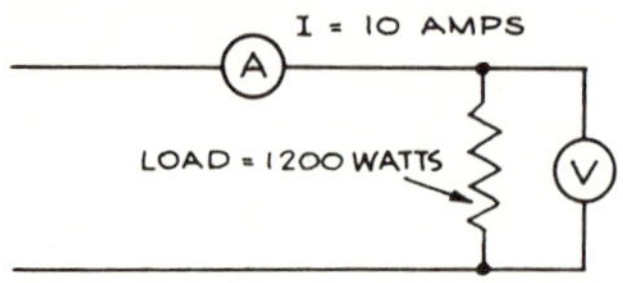

What is the voltmeter reading?

$$E = \frac{P}{I} = \frac{1200 \text{ watts}}{10 \text{ amps}} = \underline{120 \text{ volts}}$$

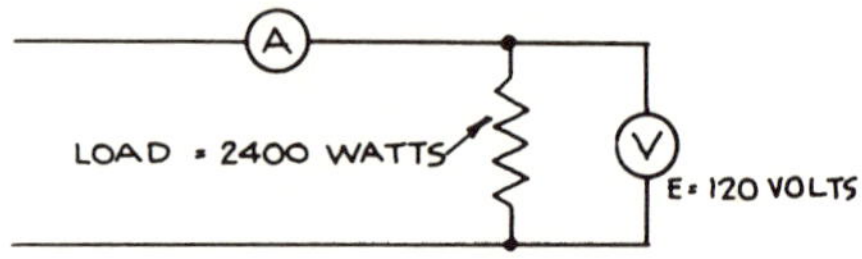

What is the ammeter reading?

$$I = \frac{P}{E} = \frac{2400 \text{ watts}}{120 \text{ volts}} = \underline{20 \text{ amps}}$$

the task. Greater accuracy is possible by using a meter range just above the value of voltage being measured.

Ammeter—This is another direct-reading instrument, designed to measure amperes of current flow. A conventional ammeter must be connected in series with the load device or circuit element through which current flow is to be measured. In the proper connection, all the current flowing through the device under test also flows through the ammeter. Another type of ammeter, the clamp-on ammeter, is magnetically connected to the circuit under measurement.

Wattmeter—This direct-reading instrument which measures power or the rate of energy expenditure is actually a combination of ammeter and voltmeter, measuring volts and amps and indicating their product, which is watts. The basic wattmeter has four terminals for connection into the circuit which is to be measured. Two of the terminals are for connection in series with the load and supply the ammeter section of the instrument. The other two terminals are for the voltmeter section and are connected across the load.

Watthour meter—This direct-reading meter is used to measure amount of electrical energy expended, the amount of electrical work done. The meter that the power company uses to determine how much energy the owner or tenant must pay for is a watthour meter. As you might guess, its internal mechanism correlates the power requirements of the loads being served and the time for which the power is applied. The product of watthours is indicated on a system of calibrated dials on the face of the meter.

Although there are many other types of meters used in today's electrical systems, the foregoing are for both dc and ac measurement.

Electric Conductors

GENERATION, transmission and utilization of electric energy depend upon the use of conductors to accomplish flow of current.

The basic electric conductor is the wire. According to standard definitions, a wire is a single, solid length of drawn metal. However, modern usage of the word "wire" includes both single, solid lengths of metal and single assemblies of a number of solid lengths of metal twisted or braided together. Thus, we have the terms "solid wire" and "stranded wire."

Particularly in the larger sizes of conductors, a single insulated or uninsulated stranded wire may be called a cable. However, the word cable is most accurately descriptive when used to indicate an assembly of two or more individually insulated wires which are twisted together, either with or without a covering over the assembly.

As discussed previously, copper is the most common material used for electric conductors. Of all of the conductive materials available, copper represents the most practical choice when such factors as conductivity, cost, strength and general adaptability to electrical purposes are considered. For instance, only silver has a higher conductivity than copper, but its high cost limits its use to special applications—such as switch contacts or fuse elements—where it offers special advantage.

Second to copper as a common conductor material is aluminum, which offers light weight and low cost as advantages to offset its somewhat lower conductivity than copper. Through recent years, use of aluminum wire and cable has grown rapidly. Other, lesser used conductive materials include: zinc, nickel, mercury, iron, tungsten and even gold.

Just as water pipe is designated according to its diameter, thereby indicating its ability to conduct quantity of water, so electric wires may be designated according to their diameters, as indication of ability to carry current. But because wires have such small diameters, it is more convenient to express the diameters in thousandths of an inch. And, further, to avoid the use of decimal points and to make reference to these diameters a simple matter, diameters of wires are given in "mils"—thousandths of an inch.

One mil is 1/1000 of an inch or

CONDUCTOR AREAS

SIZE (AWG)	AREA (CIRCULAR MILS)
18	1,624
16	2,583
14	4,107
12	6,530
10	10,380
8	16,510
6	26,250
4	41,740
3	52,640
2	66,370
1	83,690
1/0	105,500
2/0	133,100
3/0	167,800
4/0	211,600

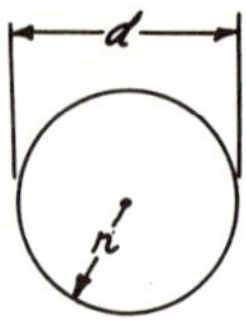

The CIRCULAR MIL area of a conductor is equal to the diameter (in mils) squared:

$$CM = d^2$$

The SQUARE MIL area of a conductor is equal to 3.1416 times the radius (in mils) squared:

$$SQ\ MILS = 3.1416\, r^2$$

1 MIL = .001 INCH
1 INCH = 1000 MILS

MILS = INCHES X 1000
INCHES = MILS X .001

1 CM = .7854 SQ MIL
1 SQ MIL = 1.2732 CM

CM = SQ MILS X 1.2732
SQ MILS = CM X .7854

$$SQ\ IN = \frac{SQ\ MILS}{1{,}000{,}000}$$

SQ MILS = SQ IN X 1,000,000

0.001 inch. There are a thousand mils to an inch. Thus a wire which has a diameter of, say, ½ inch has a diameter of 500 mils. A No. 12 wire has a diameter of 80.8 mils.

A better indication of a wire's current-carrying capacity is its cross-section area. And inasmuch as diameters are expressed in mils, areas are similarly expressed. The area of an electric wire is generally given in units known as "circular mils." A circular mil is the area of a circle having a diameter of one mil.

The cross-section area of any wire in circular mils is equal to the square of the wire diameter in mils. For instance, the circular-mil cross-section area of the No. 12 conductor mentioned above is equal to the square of the diameter in mils—80.8 x 80.8 or 6530 circular mils.

Although wires are designated according to mils of diameter and circular mils of cross-section area, and such data is commonly used in circuit calculations, the most common reference for many sizes of wire is made according to the size designations of the standard wire gages. Electric wires are generally drawn to diameter sizes which correspond to the American Wire Gage (abbreviated AWG or A.W.G.), which is also called the Brown and Sharpe (B. & S.) wire gage.

According to this gage, each size of wire is given a number ranging

RESISTIVITY OF TYPICAL METALS
(OHMS PER CIRCULAR-MIL-FT AT 20°C)

SILVER	9.9
COPPER (ANNEALED)	10.37
GOLD	14.7
ALUMINUM	17
TUNGSTEN	33
BRASS (ANNEALED)	42
NICKEL	47
PLATINUM	60.2
CONSTANTAN	295
NICHROME	602

DIAMETER OF SOLID CONDUCTOR IN MILS = $\sqrt{\text{AREA IN CIRCULAR MILS}}$

(This does not apply to stranded conductors (No. 6 and larger), since the area in CM is the area of the copper (or aluminum) only, not the overall area of the conductor.)

$$RESISTANCE = \frac{RESISTIVITY \times LENGTH\ IN\ FT}{AREA\ IN\ CM}$$

EXAMPLES:

1. What is the diameter of a No. 10 conductor?

NO. 10 AREA = 10,380 CM

DIA. = $\sqrt{10,380}$ = 102 MILS

= 0.102 IN.

2. What is the sq in. area of a No. 8 conductor?

NO. 8 AREA = 16,510 CM

16,510 × 0.7854 = 13,000 SQ MILS

$\frac{13,000}{1,000,000}$ = .013 SQ IN.

3. What is the cross-section area of a bus bar equivalent in area to a 500MCM conductor?

AREA = 500,000 CM

500,000 × .7854 = 393,000 SQ MILS

$\frac{393,000}{1,000,000}$ = 0.393 SQ.IN.

(Note: The National Electrical Code gives the area as 0.520 sq in. This is not the area of the copper or aluminum, but the area of a circle with a diameter equal to the overall diameter of the stranded 500MCM conductor.)

4. How much current would flow in the circuit shown at right if the conductors were (a) copper, (b) brass, (c) Nichrome?

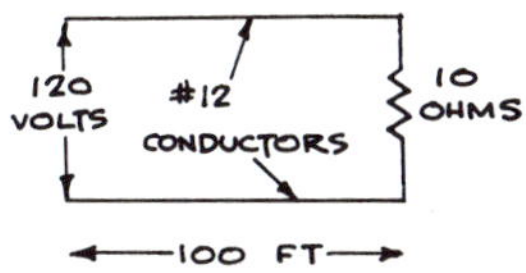

(a) CONDUCTOR RESISTANCE = $\frac{10.37 \times 200}{6530}$ = 0.318 OHM

TOTAL RESISTANCE = 0.318 + 10 = 10.318 OHM

CURRENT = $\frac{\text{VOLTS}}{\text{OHMS}}$ = $\frac{120}{10.318}$ = 11.6 AMPS

(b) CONDUCTOR RESISTANCE = 1.29 OHMS

TOTAL RESISTANCE = 11.29 OHMS

CURRENT = 10.65 AMPS

(c) CONDUCTOR RESISTANCE = 18.4 OHMS

TOTAL RESISTANCE = 28.4 OHMS

CURRENT = 4.23 AMPS

from No. 40 the smallest size, up through No. 1, No. 0 (read as "one naught") No. 00 or 2/0 ("two naught"), No. 000 or 3/0 ("three naught") and, the largest gage size, No. 0000 or 4/0 ("four naught"). Wire sizes larger than No. 4/0 are designated according to their circular mil size, e.g., 500,000 circular mils. And in these larger sizes, the letters "MCM" are used to abbreviate thousand circular mils, so that 500,000 circular mils becomes 500MCM.

It should be noted that in the gage sizes the higher numbers are the smaller wires. Thus No. 12 is a smaller wire than, say, No. 4 and No. 1 is smaller than No. 1/0. And, of course, the ability of conductors to carry current increases with increase in size. Such current ratings for wires may be based on the capability of the insulation on the wire to withstand the heat developed by the I^2R energy loss in the wire or it may be limited by the voltage drop in the wire.

As described previously, electron flow through conductive materials depends upon the relative "looseness" of the atomic structure of the material. When the structure is very loose so that there is very low obstruction to current flow, the material is said to have "low resistivity" or, describing the reciprocal "high conductivity."

If the structure is relatively tight, the material has "high resistivity" or "low conductivity." For any conductor, the resistance—in ohms—of a given length depends upon the resistivity of the material, the length of the wire and the cross-section area of the wire in accordance with the formula—

$$R = \rho \frac{L}{A}$$

VOLTAGE DROP

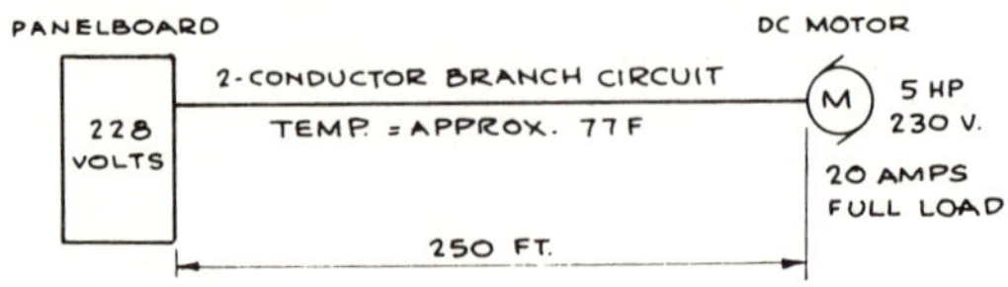

PROBLEM:

Speed and torque requirements demand that the terminal voltage of the motor shown be not less than 223 volts at rated current. Utility voltage variations produce a minimum of 228 volts at the panel. What size branch circuit conductors are required?

SOLUTION:

The National Electrical Code, Sec. 430-22, requires that branch circuit conductors supplying a single motor have a carrying capacity not less than 125% of the motor full-load current rating.

1.25 x 20 AMPS = 25 AMPS.

This calls for a #10 conductor.

Maximum allowable voltage drop
= 228 - 223 = 5 VOLTS.

From table, resistance of #10 wire
= 1.018 OHM / 1000 FT.
= .509 OHM FOR 500 FT.

Voltage drop = 0.509 OHM x 20 AMPS
= 10.18 VOLTS.

This is more than twice the allowable voltage drop; therefore the conductor resistance is twice as high as permissible.

From table, #6 wire has a resistance of 0.410 ohm /1000 ft or .205 ohm for 500 ft.

Voltage drop = .205 OHM x 20 AMPS
= 4.10 VOLTS.

Terminal voltage = 228 - 4.10
= 223.9 VOLTS.

Since this exceeds the minimum requirement of 223 volts, #6 conductors may be used.

DC RESISTANCE OF BARE (UNTINNED) COPPER CONDUCTORS

AWG	OHMS / 1000 FT AT 25C (77 F)
14	2.575
12	1.619
10	1.018
8	.641
6	.410
4	.259
3	.203
2	.162
1	.129
0	.102
00	.0811
000	.0642
0000	.0509

where, R = the resistance of the conductor in ohms. L = the length of the wire, A = the area of the wire cross section and ρ (the Greek letter "r", pronounced *row*) = the resistivity. The unit of resistivity is often expressed in ohms per circular-mil-foot. This is the resistance of a wire of the given material which is one circular mil in cross section and one foot long.

Resistivities of various metals are shown in the table on the opposite page. Example: Find the resistance of a 100-foot length of copper wire with a diameter of 5 mils at 20° C.

The resistivity of copper in this case is 10.37 ohms per circular-mil-foot. The cross-section area is $(5)^2$ or 25 circular mils. The total resistance, then, is—

R = 10.37 x 100 feet/25 circular mils or 41.48 ohms

The example illustrates that resistance increases with length of the wire but decreases with increase of wire cross section.

IN ALL phases of electrical practice, reference to the characteristics of electrical conductors is continually made and calculations based on such characteristics are common.

For this reason, there are some basic relationships among the various gage sizes of wires which should be known:

1. In the American Wire Gage, there is a constant ratio between the diameter of any wire and the diameter of the next smaller wire. This ratio is approx. 1.12/1. Then, since the area in circular mils is equal to the square of the diameter in mils, the ratio of successive cross section areas is $(1.12)^2$ or 1.26/1. Thus, for example, the ratio of cross section areas of No. 2/0 and No. 1/0 conductors is 133,100 circular mils divided by 105,500 circular mils—which is equal to 1.26. The ratio of any two successive cross-section areas in the AWG table is always 1.26.

2. An increase of three gage numbers doubles the cross-section area of the conductor and also doubles the weight for the same conductor material. For example, No. 1/0 has a cross section of 105,500 circular mils. Three gage sizes larger is No. 4/0, which has a cross-section area of 211,600 circular mils or double that of No. 1/0.

3. An increase of three gage numbers halves the conductor resistance. Thus, from the previous paragraph, a No. 4/0 conductor has a dc resistance of 0.0509 ohms/1000 feet. The No. 1/0 conductor has a dc resistance of 0.102 ohms/1000 feet—twice that of the No. 4/0 conductor.

4. An increase of 10 gage numbers increases cross-section area and weight by a factor of 10 and decreases resistance by a factor of 10. Thus, a No. 1/0 conductor, which is 10 gage sizes larger than No. 10, has 10 times the area and weight of the No. 10 and 1/10 of the resistance of the No. 10.

5. The ratio of 1.26 between successive cross-section areas may also be applied to successive resistance values, observing that resistance decreases with increase in cross-section area. For example, just as the cross-section area of a No. 3 conductor is 1.26 times the cross-section area of a No. 4, the resistance of the No. 4 is 1.26 times the resistance of the No. 3. The resistance of a No. 3 is 0.205 ohms/1000 feet. The resistance of a No. 4 is 0.205 x 1.26 or 0.259 ohms/1000 feet.

6. All of the foregoing data can be used to arrive at almost any characteristic of any size of conductor by applying it to the following easily-memorized data on No. 10 copper conductor at normal room temperature—

No. 10 has a diameter of 0.1 inches or 100 mils.
No. 10 has a cross-section area of $(100)^2$ or 10,000 circular mils.
No. 10 has a dc resistance of approx. 1 ohm per 1000 feet.
No. 10 weighs 32 pounds per 1000 feet.

For ready reference on the properties of conductors, extensive data is given in a number of tables in Chapter 9 of the National Electrical Code.

One of the most common design considerations involving the resistance characteristics of conductors is that of voltage drop. As stated previously, there must be a certain amount of voltage drop in the conductors which make up modern feeders and branch circuits because all conductors have resistance and voltage must be used to push current through them. Of course, the amount of voltage used to get the current through the conductors should be kept to a relatively small percentage (1%, 2% or maybe

EFFECT OF TEMPERATURE ON RESISTANCE

The resistance R_x of copper wire at any ambient temperature t_x may be found using the formula

$$R_x = R\left(\frac{234.5 + t_x}{234.5 + t}\right)$$

where R = known resistance at temperature t in degrees C.

EXAMPLE:
What is the resistance of #6 wire at a temperature of 45C?

SOLUTION:
From table, R = 0.41 at 25C.

$$R_x = .41\left(\frac{234.5 + 45}{234.5 + 25}\right)$$

$$= .41 \times \frac{279.5}{259.5}$$

$$= .41 \times 1.077$$

$$= .442 \text{ OHM}/1000 \text{ FT.}$$

up to 10%) of the voltage impressed at the supply end of the circuit. But some voltage must be dropped and the amount of this voltage depends upon resistance—either resistance alone in the case of direct current circuits or resistance and reactance (an effect opposing current flow in circuits with alternating current).

As will be described later, there is a difference between dc resistance and ac resistance, with ac resistance being higher for a given size of conductor only in the larger sizes of conductors—conductors No. 2 and larger.

As pointed out earlier, some conductors are made up of a single solid strand of conductive material and some are made up of a number of strands twisted together. The stranded conductor provides a much more flexible conductor assembly for easier handling, bending and installing.

For this reason, the stranded assembly is used for the LARGER sizes of conductors—No. 6 and larger, although stranded assemblies are also used for smaller sizes for such applications as lighting fixture wiring and other equipment wiring. The NE Code, for instance, requires that conductors No. 6 and larger when used in raceways must be of the stranded type.

In practice, the safe current-carrying capacities of electrical conductors is based on the insulation of the conductors. Current limits are those which would produce heating that would damage or destroy the insulation. Manufacturers of wires and cables make available complete tables of conductor current ratings.

Again, the NE Code presents maximum continuous current ratings for conductors with various insulations. However, it is common to find that design of circuits for low voltage drop and low power losses in the lines requires that loading of conductors (amount of current flowing) be kept considerably below the current levels which establish the temperature limit of application.

An important characteristic of the resistance of conductor materials is its relation to temperature. The resistance of the usual conductor materials —copper and aluminum—increases with increases in the temperature of the material. There are some conductive materials—carbon and conductive solutions—which decrease in resistance as temperature rises. Still other materials show very little resistance change with temperature variation. Resistance values—as given in most tables and texts—are those which obtain for usual room temperatures—from 68 to 77 degrees F (20 to 25 degrees C). The temperature coefficient of resistance of a conductor material is the percent change in resistance that will occur for each degree C change in temperature, based on the resistance at some initial temperature in degrees C. A commonly used temperature base for establishing the coefficient is 20 degrees C.

Basic Circuit Study

ALL of the foregoing principles and characteristics of electric current flow apply to direct-current and alternating-current circuits.

There are additional principles and characteristics which apply only to alternating-current circuits. These will be covered later. So far, the theory of electricity can be correlated to practical circuits and application as follows:

Voltage Source—For the vast majority of buildings, the source of voltage which powers the entire interior distribution system is the service from the electric utility company. This is simply connection to the end of a feeder from the utility company's generator.

In some cases, the source of voltage for a building may be a generator right on the premises. And there are many cases in commercial and industrial application where a segment of a building's system is supplied from a small generator or from a dc source such as a bank of batteries or a rectifier. This latter case covers special needs for voltage different from the basic system voltage or needs for emergency sources of power in the event of failure of normal power.

Conductors — Connection from the service conductors of a building to the various load devices is made by a simple or complex network—depending upon the size and type of building and loads—of wires and cables of varying sizes and constructions. From the utility terminals to the main switch or main switchboard, the conductors are the service entrance conductors.

From the main switchboard to each panelboard in the building, two or more conductors provide a circuit known as a feeder, to deliver power to the circuits of the panel. A feeder from a panelboard to another panelboard is commonly called a subfeeder. Then conductors which run from a panelboard to motors, lighting fixtures or plug outlets make up the branch circuits of the system.

Each of the conductors throughout the entire system has sufficient cross-section area to carry the current impressed upon it—without producing heat of I^2R which might damage the insulation on the conductor and without producing so much voltage drop by current flow through the resistance of the conductor (drop I R) that the voltage delivered to the load devices is objectionably low. In practice, voltage drops are generally held to no more than 5% of the nominal system voltage. A widely used approach to voltage drop holds the drop to about

1% for supply to lighting loads and 3% for circuits to motor loads.

Load Devices—These are the utilization devices for which the entire electrical system is installed in the building. Each lighting and heating device is designed to operate at a nameplate voltage which will produce a certain current flow through the device, depending upon the resistance of the device. Fluorescent and mercury vapor lamps use coil type ballasts which oppose alternating current flow by the resistance of the wire in the coil and by an alternating current characteristic known as inductive reactance. Motors, both ac and dc, have resistance to oppose current flow, and ac motors have inductive reactance.

Three-Wire Circuits

The basic circuit referred to thus far has been the two-wire circuit to a lamp or other load device. And it has been shown that this same circuit may be used to supply a number of lamps connected in parallel—the usual method of load connection in modern electrical systems. Then, too, the same circuit of only two wires can be used to provide the feeder to the panelboard from which the two-wire branch circuits are derived. And the basic two-wire circuit used over the past years has been the 120-volt circuit to supply the very many load devices designed for that voltage.

The three-wire circuit was developed to reduce the problems of voltage drop and power loss in the lines, while still providing for supply to 120-volt circuits. In the two-wire circuit, increase in the size of the circuit conductors reduces the resistance of the line and thereby decreases voltage drop and power losses. But this is the only thing that can be done for improving two-wire circuits. However, the use of higher voltage offers the chance to deliver a given amount of power at lower current than would be required for the same power at 120 volts, i. e., because power equals E x I, the current is less at the higher voltage. And, as a result, the voltage drop and power loss in the circuit conductors will be lower.

But use of the higher voltage for a two-wire circuit does not meet the need for supplying 120-volt loads. The three-wire circuit, which consists of two wires with 240 volts between them and 120 volts between each of these wires and a third wire, provides the advantages of reduced voltage drop and power loss, with ability to serve 240-volt load devices and without sacrifice in the ability to supply 120-volt loads. Such a circuit is commonly described as a 120/240-volt, three-wire circuit and may be used for direct current or alternating current.

In the three-wire circuit, the conductor which forms a 120-volt circuit with either of the other two wires is called a neutral conductor and is generally required by electrical codes to be grounded (this matter of grounding will be discussed later). When the load resistance on one of the 120-volt circuits is equal to the load resistance on the other 120-volt circuit, the neutral conductor carries no current and the other two wires carry the same current. In effect, the two load resistances are connected in series across the 240-volt circuit of the non-neutral conductors. The total power supplied is equal to 240 times the current flowing.

The voltage drop and line power loss under these conditions are the same as they would be for a 120-volt circuit of the same size wires, carrying the same current and supplying only half the power. Thus for a given load on each of the 120-volt circuits, the voltage drop and power loss for each 120-volt circuit are only one-half of what they would be for a simple two-wire 120-volt circuit serving the same load as each of the other two 120-volt circuits. And when the load resistance on each of the two 120-volt circuits of a 120/240-volt three-wire circuit is not the same, the neutral wire carries the difference between the currents flowing in the non-neutrals.

3-WIRE CIRCUITS

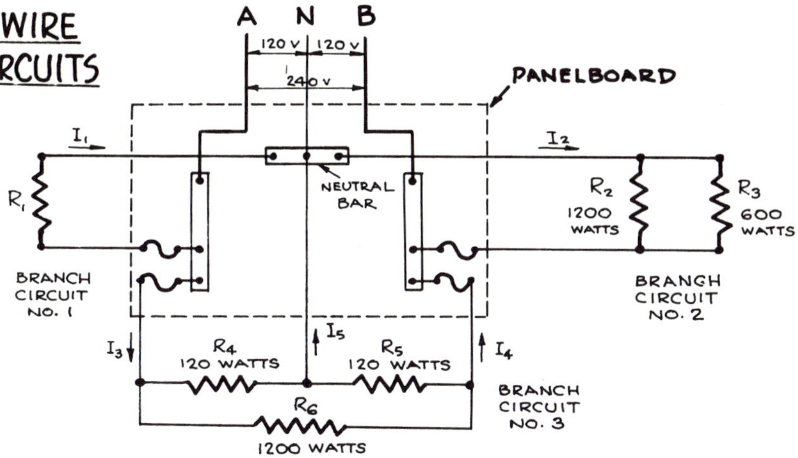

Illustrated is a panelboard fed by a 3-wire, 120/240-volt feeder. Three branch circuits are shown; Circuits No. 1 and 2 are 2-wire circuits; No. 3 is a 3-wire circuit.

(1) *If the current I_1 in Circuit No. 1 is 12 amps, what is the wattage of load R_1? (Neglect voltage drop.)*

R_1 = 120 VOLTS × 12 AMPS = 1440 WATTS

(2) *What is the current I_2 in Circuit No. 2?*

CURRENT IN R_2 = $\frac{\text{1200 WATTS}}{\text{120 VOLTS}}$ = 10 AMPS

CURRENT IN R_3 = $\frac{\text{600 WATTS}}{\text{120 VOLTS}}$ = 5 AMPS

I_2 = 10 + 5 = 15 AMPS

(3) *What current flows in feeder conductors A, B and N if Circuits No. 1 and 2 are energized and Circuit No. 3 deenergized?*

CURRENT IN "A" = I_1 = 10 AMPS

CURRENT IN "B" = I_2 = 15 AMPS

THESE TWO CURRENTS OPPOSE EACH OTHER IN THE NEUTRAL, AS SHOWN IN THE FOLLOWING DIAGRAM; 10 AMPS FLOW IN ONE DIRECTION, 15 AMPS IN THE OTHER. THE NET CURRENT IN "N" IS THUS 15 − 10 OR 5 AMPS.

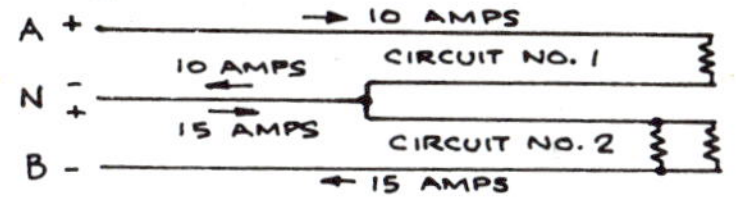

(4) *In Circuit No. 3, R_4 and R_5 are 120-volt loads; R_6 is a 240-volt load. Find I_3, I_4 and I_5.*

CURRENT IN R_4 & R_5 = $\frac{\text{120 WATTS}}{\text{120 VOLTS}}$ = 1 AMP

CURRENT IN R_6 = $\frac{\text{1200 WATTS}}{\text{240 VOLTS}}$ = 5 AMPS

I_3 = 1+5 = 6 AMPS = I_4 I_5 = 0, SINCE I_3 = I_4

(5) *If R_4 is turned off, find currents I_3, I_4 and I_5.*

CURRENT IN R_4 = 0

CURRENT IN R_5 = 1 AMP

CURRENT IN R_6 = 5 AMPS

I_3 = 0 + 5 = 5 AMPS I_4 = 1 + 5 = 6 AMPS

I_5 = 6 − 5 = 1 AMP

Magnetism

MAGNETISM is a physical phenomenon which plays an essential role in the distribution and utilization of electrical energy.

Although we can generate light and heat by simply passing either direct or alternating current through a resistance—such as the element of a toaster or the filament of an incandescent lamp, rotating electrical equipment like motors and generators and other devices such as relays, transformers, induction heating equipment —depend upon magnetism.

Fundamentals

Centuries ago, it was discovered that a certain iron ore found in the earth had the ability to attract and hold to itself pieces of iron. Because this ore was plentiful in Magnesia, a province in Asia Minor, it was given the name "magnetite." From this, the name "magnet" is used to describe any body attracting iron or steel.

Although natural magnets as found in the earth have no useful application, artificial magnets are readily manufactured from iron, steel or various alloys, and these have widespread commercial value. A piece of steel, for instance, can be made into a magnet by repetitively stroking it with a body which is already a magnet. But the most common way of magnetizing such a piece of steel is to place it within a coil of wire and pass a heavy current through the wire. The flow of current sets up a magnetizing force which orients the molecular structure of the steel, giving it the character of a magnet.

Magnets are commonly classified according to their ability to retain the character of a magnet. Those which retain the strength of their magnetism indefinitely are called **permanent magnets.** In this category are hardened steel and a number of scientifically developed alloys of nickel and cobalt. Those magnets which owe the strength of their magnetism to an external magnetizing force (such as the current-carrying coil of wire on the core

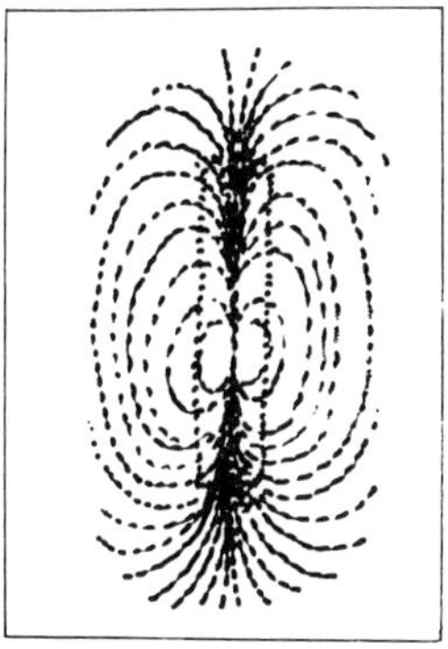

LINES OF FORCE *are demonstrated by iron filings sprinkled on sheet of paper over bar magnet. Filings outline magnet and lines of force. Points of greatest concentration are points of greatest magnetic attraction.*

MAGNETIZATION *of a piece of steel, such as a screwdriver blade, may best be accomplished by placing it inside a coil of wire carrying a heavy dc current and shutting off current while blade is inside.*

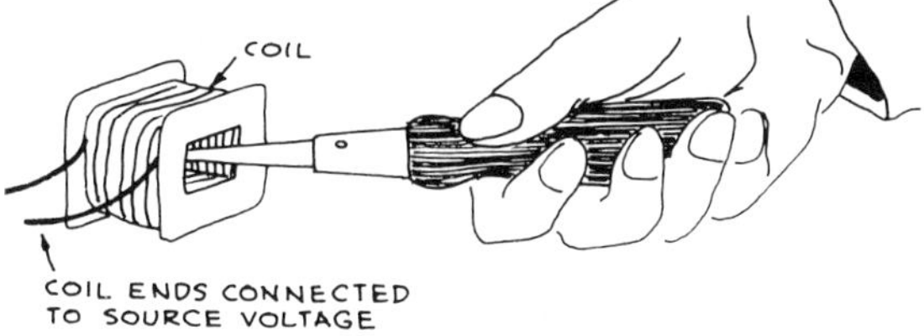

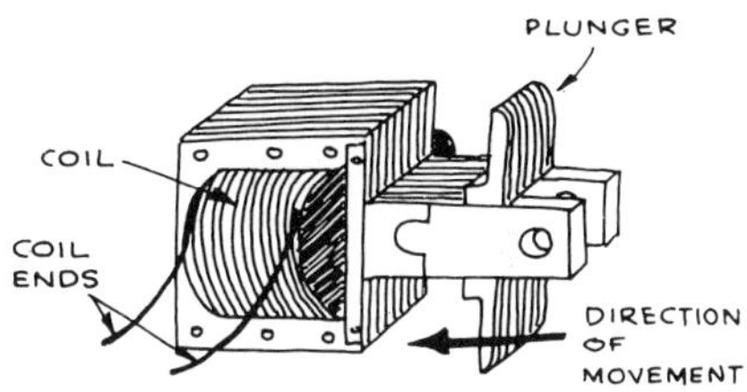

SOLENOID COIL *carrying current attracts steel plunger, pulling it into coil. Considerable manual force is necessary to withdraw plunger. When current is shut off, plunger may be withdrawn easily; coil loses its magnetism. This is an example of a "temporary" magnet.*

of an electro-magnet) and which lose almost all of their magnetism when the external force is removed are referred to as **temporary magnets.**

Two terms used in connection with magnets are: permeability and retentivity. **Permeability** refers to the relative ease with which a material can be magnetized, the ease with which its molecular structure can be oriented to that of a magnet. Soft iron, as an example, is relatively easy to magnetize and is said to have high permeability. Steel, on the other hand, is hard to magnetize and is said to have lower permeability than soft iron.

Retentivity is a specific term referring to the ability of any magnet to retain its magnetic character, with permanent magnets having high retentivity and temporary magnets having low or poor retentivity. In the case of temporary magnets (soft iron) that small amount of magnetism which is retained after the external magnetizing force is removed is called **residual magnetism.**

The term **magnetic materials** is used to refer to those materials which are attracted or repelled by a magnet and which can be magnetized themselves. Iron and steel are the most common of such materials.

Magnetic Field

Magnets can be made in a wide variety of shapes and sizes, from any one of the various magnetic materials available. **But there are certain characteristics common to all magnets:**

1. If a straight bar of magnetized steel or a horseshoe-shaped magnet or any other shape is dipped into a pile of iron filings, it will be found that the filings cling in large numbers to the ends of the magnet and only a relatively few filings attach themselves to the intermediate part of the magnet. The area at each end of a magnet, where the attractive force is the greatest, is called a **pole.**

2. The poles of any magnet are the points at which certain forces are concentrated. This force takes the form of lines of energy and is specifically referred to as **lines of force.** The existence of these lines of force can be revealed in two ways:

A. If a piece of white paper is placed over a bar magnet and iron filings are sprinkled evenly over the paper, the filings will be seen to take the form of actual lines running, external to the magnet, from one pole to the other in curving paths. The filings are actually under the influence of the lines of force.

B. If the same bar magnet is placed on a flat surface, a small compass can be used to study the presence of the lines of magnetic force. By placing the compass in various positions along the length of the bar magnet, it will be seen that the needle takes different positions. The North pole of the needle will always point toward the South pole of the bar magnet, and the needle will always line itself up with the direction of the lines of force at the particular position of the compass. The needle points away from the North pole of the magnet. By plotting the various positions of the compass needle, the lines of force will be indicated on both sides of the magnet.

3. From the above, lines of magnetic force have definite direction at all points in the space surrounding a magnet and they act along curved lines.

4. The strength of a magnet is concentrated at the two poles. This accounts for the high strength of attraction for iron or steel at the poles.

5. The total quantity of magnetic lines of force is commonly called **magnetic flux.** And, again, flux is always from North to South external to the magnet and from South to North internal to the magnet. Lines of force are continuous and never cross.

6. Although magnetic lines of force are really imaginary, they provide a helpful, working concept. And the lines of force exist in the full 360° of space around the axis of the magnet.

7. The flow of magnetic flux can be roughly compared to the flow of electric current. The magnet is the source of pressure (comparable to the voltage source) and the flux finds more or less opposition to its presence. This opposition is known as the **reluctance** (similar to resistance in electrical circuits) of the material. Air has very high reluctance and iron and steel have lower reluctance. And the lower the reluctance for the lines of force from a given magnet, the higher the flux density.

IF A BAR MAGNET is suspended freely at the end of a length of string and the North pole of a second, hand-held, bar magnet is brought slowly toward the North pole of the suspended magnet, it will be readily seen that a force of repulsion exists between the two North poles.

The North pole of the suspended magnet will move away from the other North pole. And the same repulsion exists between two South poles. Then just the opposite effect will be achieved if the North pole of the hand-held magnet is slowly moved toward the South pole of the suspended magnet. In this case, there is an attractive force between the two poles and they move quickly together. These two actions illustrate the **first laws of magnetism:**

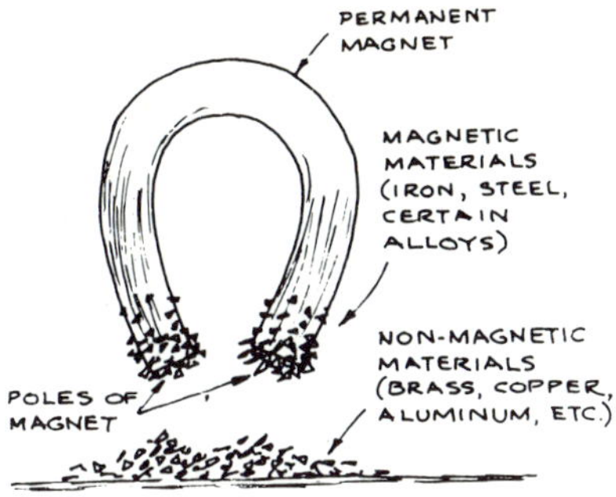

PERMEABILITY of metals may be demonstrated by dipping permanent magnet into pile of metal fragments. Those with high permeability (magnetic) will adhere readily to the magnet; those with low permeability (non-magnetic) will not. This principle is employed in magnetic separators used in industry.

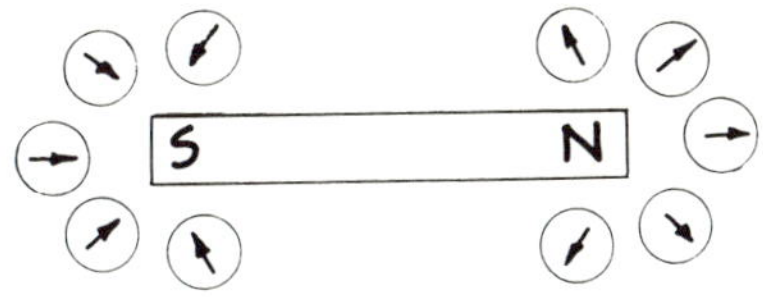

COMPASS, moved about a bar magnet, will always point away from a North magnetic pole, aligning itself with the bar magnet's lines of force. The compass needle itself is a small bar magnet with a North and a South pole; its South pole is attracted to the larger magnet's North pole and vice versa.

1. Like magnetic poles repel each other.

2. Unlike magnetic poles attract each other.

It has been noted that the needle of a compass (a small magnet, itself) always points North and South. This occurs because the North pole of the little bar magnet in the compass is attracted toward the earth's geographic North pole which is actually a South magnetic pole.

The character of magnetism is such that if a bar magnet is cut into two pieces, each piece is a magnet in the same sense that the whole bar was. And further cutting of these pieces into any number of pieces will yield that number of little magnets. One theory of magnetism holds that in any magnet even the molecules act as magnets. Accordingly, the theory says that these molecules in iron or steel may be arranged without any orientation among them and such a piece of iron or steel is unmagnetized. However, if the same piece of iron or steel is subjected to a strong magnetic field, the magnetic molecules will be forced into a definite pattern of orientation. Such orientation of the molecules makes a magnet out of the piece of iron or steel. In it, all of the molecular North poles point one way and all of the molecular South poles point in the opposite direction. Removal of the magnetizing force from a piece of soft iron will result in the magnetic molecules resuming their original position. But the molecules of hardened steel will retain their orientation even after the magnetizing force is removed, becoming a permanent magnet.

Electromagnetism

In 1819, a scientist named Oersted discovered a relation between magnetism and electric current. He found that an electric current flowing through a conductor produced a magnetic field about the conductor. When a small compass was brought close to the wire carrying current, the needle of the compass was deflected at right angles to the wire. If a current-carrying conductor is passed through a sheet of cardboard, the nature of the magnetic field around the conductor may be studied by placing the small compass on the cardboard and moving it in circles around the wire.

It will be noted that the needle keeps changing position slightly as it is moved around the conductor, indicating that the magnetic field exists in concentric circles around the conductor and that it has definite polarity, i.e., the needle points in one direction with the current flowing in a given direction and then points in the opposite direction when the direction of current flow is reversed in the wire.

Another way of detecting the presence of the magnetic field around the conductor is by sprinkling iron filings on the cardboard and noting the pattern of concentric circles formed by the filings. The iron filings become magnetized by the field around the conductor, and the interaction of the two sets of magnetic force lines the filings up in circles, like a large number of compass needles.

It can be seen from the two experiments that the magnetic field around the wire extends some distance out from the wire, and the strength of the field is greatest close to the wire and decreases with distance from the wire. And another noteworthy fact is that the strength of the field can be increased by increasing the current flowing through the wire and decreased by decreasing current flowing through the wire. This can be noticed by watching the stronger current make larger, more concentrated rings of the iron filings than weaker current is able to do.

All of the foregoing forms the basis for the so-called "Right Hand Rule." This is a handy, convenient means of determining the relationship between the flow of electric current in a wire and the direction of the magnetic lines of force around the magnet. It works this way: Grasp the current carrying wire in the right hand, wrapping the four fingers around the wire and extending the thumb along the wire. If the thumb points along the wire in the direction of current flow, the fingers will be pointing in the direction of the lines of force around the conductor.

Although the magnetism is relatively weak at any point along the length of a wire, the magnetic effect can be strengthened by winding the wire into a coil. The total magnetic force of the length of the wire is thereby concentrated in a much smaller space and is, as a result, more intense. The character of the magnetic field of a coil can best be described by considering the simple condition of two wires placed side-by-side and carrying current in the same direction.

As shown in the sketches, the lines of force between the conductors oppose each other, keeping the magnetic field around the two conductors —or around three or four or however many conductors are placed together. As a result, the lines of force associated with a coil of wire flow in one direction through the core of the coil and flow in the opposite direction around the outside of the coil where they return to the other end of the core. The end from which the flux emerges is the North pole of the electromagnetic coil. And since all of the turns of the coil have current flowing in the same direction through them, their fields add up and polarity of the coil can be determined by a variation on the "Right Hand Rule."

Grasp the coil in the right hand with the fingers wrapped around the coil in such a way that the fingers point in the direction of the current flow through the turns of the wire. The thumb then points toward the end of the coil which is the North pole of the electromagnet.

THE ABILITY of an electromagnetic coil to produce magnetic lines of force, or magnetic flux, is measured in terms of **magnetomotive force**, which is abbreviated MMF.

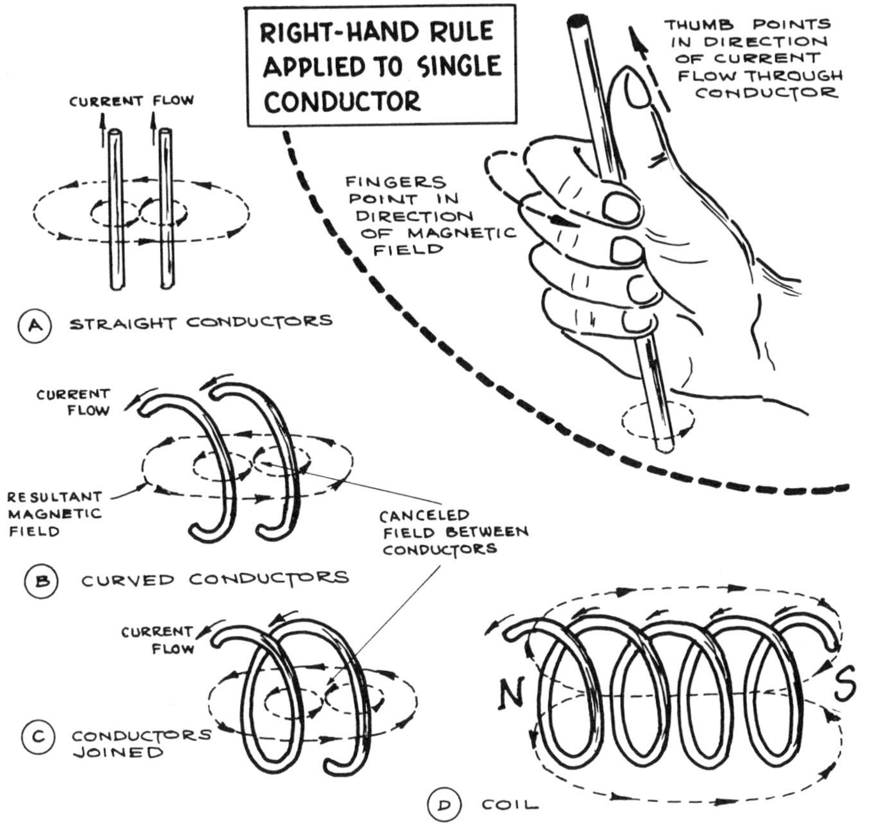

MAGNETIC FIELD OF A COIL

Right-hand rule applied to each of two adjacent conductors (A) shows lines of force between wires cancel each other, with resultant field circling both conductors. Effect is unchanged with curved wires (B) or joined as in (C). Additional wire forms a coil (D). We now have an electromagnet, with north and south poles. Lines of force travel north to south outside coil; south to north inside coil.

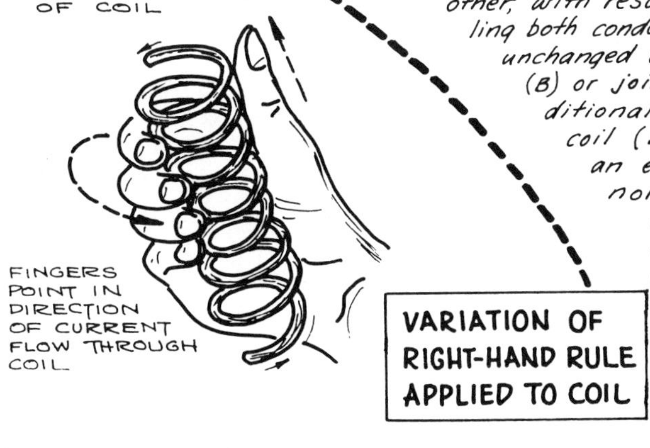

ELECTROMAGNETIC INDUCTION,

fundamental principle of the generator, may be demonstrated as shown. If conductor is moved up or down perpendicular to the magnetic field, sensitive dc voltmeter will show induced potential across ends of conductor as it cuts lines of force. Right hand, with fingers extended at right angles, affords means of determining direction of lines of force, induced emf, or conductor movement when two of the three are known.

Meter will indicate zero voltage with conductor at rest in magnetic field.

Moving conductor up through lines of force produces meter reading in one direction.

Reversing direction of conductor movement reverses meter indication.

Using a stronger magnet or increasing conductor speed increases magnitude of induced voltage.

MAGNETOMOTIVE FORCE

of a coil is measured in ampere-turns:

AMPERE-TURNS = AMPERES X TURNS

1 AMP THRU 1 TURN = 1 AMPERE-TURN

1 AMP THRU 2 TURNS = 2 AMPERE-TURNS

10 AMPS THRU 3 TURNS = 30 AMPERE-TURNS

Magnetomotive force in a magnetic circuit produces lines of force, just as electromotive force in an electric circuit produces current.

In the magnetic circuit, magnetomotive force can be considered to be the pressure which produces the magnetic lines of force through the reluctance of the magnetic path. MMF is similar to EMF, or voltage, which produces current flow through the resistance of an electric circuit.

The MMF of any coil depends upon the current flowing through the coil and the number of turns in the coil. As either or both of these increases, the magnetomotive force of the coil increases and more flux will be produced through the given magnetic circuit. The product of the current through the coil and the number of turns in the coil is referred to as the **ampere-turns** of the coil and is taken as a measure of MMF.

From the foregoing, it is clear that the strength of the magnetic field produced by an electromagnetic coil through a given magnetic path depends upon the ampere-turns of MMF just as electric current through a given circuit resistance depends upon the voltage pushing it. But, just as current through a circuit can be increased by reducing the resistance, so the flux produced by an electromagnetic coil can be increased by reducing the reluctance of the path taken by the flux. Inserting a bar of soft iron in a current-carrying coil of wire, for instance, produces a great reduction in the reluctance of the magnetic path. Magnetic flux finds it much easier to exist in magnetic materials than in air or other non-magnetic materials. As a result, a tremendous increase in flux density is produced when the iron bar replaces air as the core of the coil.

Electromagnets, consisting of coils with iron cores, find extensive practical applications in a wide range of electrical devices and machines, in addition to direct use as lifting magnets.

An important characteristic of electromagnetic flux in magnetic core materials is saturation. When the flux produced by a coil around an iron core continues to increase with increase in current, the core contains more and more lines of force. But a point is reached at which increase in current through the coil—i.e., increase in MMF—no longer produces an increase in lines of magnetic flux. At that point, the core is said to be saturated. Different magnetic materials have different points of saturation, for given cross-section area of cores, in accordance with their differing permeabilities.

Further evidence of the relation between electric current and magnetic flux is obtained by an experiment which shows that an electromotive force (a voltage) can be produced in a conductor by moving the conductor through a magnetic field in such a way that the conductor cuts the lines of force. This is the phenomenon of electromagnetic induction, discovered by Faraday in 1831 and since incorporated as the basic operating principle of such important devices as the electric generator and the transformer.

The basic experiment on induction is shown below.

A conductor, with a voltmeter connected across its ends, is moved through the magnetic field between the poles of a horseshoe magnet in such a way that the conductor is perpendicular to the lines of force and cuts through them. The voltmeter will indicate that a voltage of a definite polarity is produced between the ends of the conductor during the time it is cutting through the flux. As soon as the motion of the conductor stops, the voltage drops to zero. Then if the conductor is moved through the flux in the opposite direction, a voltage is again produced,

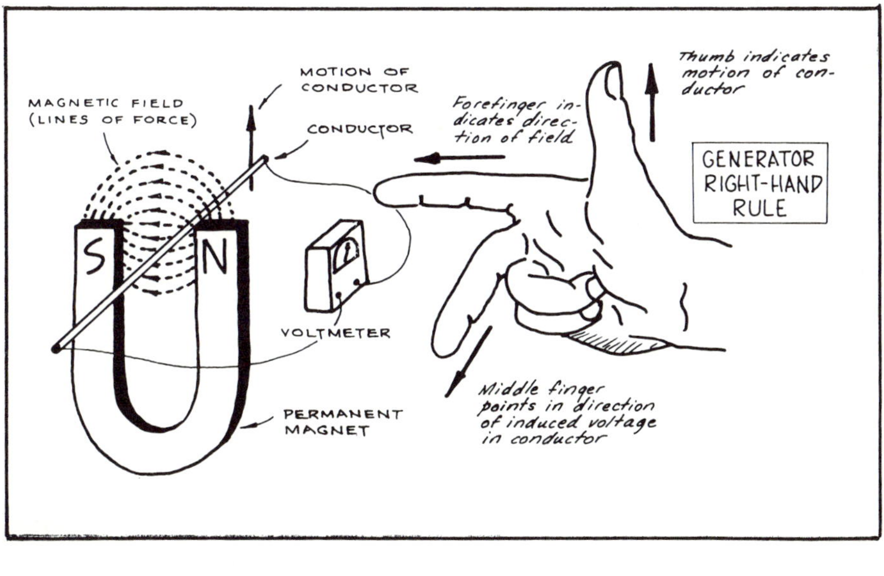

but it has polarity opposite to that produced in the first case. Another way of demonstrating the same effect is to fix the position of the conductor and move the magnet so that the flux between its poles cuts the fixed conductor. Again, the voltmeter will indicate that a voltage has been produced. And again, voltage of opposite polarity can be produced by moving the magnet so that the flux cuts the conductor in the opposite direction.

The production of a voltage by relative motion between a current-carrying conductor and a magnetic field is called **electromagnetic induction.** The voltage produced in the conductor is said to be an **induced voltage**. If this induced voltage is connected to a closed circuit, a current will flow, and such current is said to be **induced current.**

The most important application of relative motion between conductor and magnetic field is made in electric generators. In a dc generator, fixed electromagnets are arranged in a cylindrical housing to set up the magnetic field and a very many conductors in the form of a coil are rotated on a core within the magnetic field in such a way that these conductors are continually cutting through the magnetic flux. As a result, voltage is produced in each of the conductors, and since the conductors are all in series in the coil, the induced voltages add together to produce the output voltage of the generator.

The same principle is used in ac generators, although it is common to find the coil of conductors fixed and the electromagnets rotated so the field cuts the conductors.

The value of EMF produced by moving a conductor through a magnetic field can be increased by adding more conductors (such as winding a long conductor into a coil so that each turn of the coil has a voltage induced in it and the voltages of the turns are added to each other to provide a higher voltage). Another way to increase induced voltage is to increase the speed at which the conductors cut through the magnetic flux. This can be done by increasing the speed of rotation of the core on which the conductors are wound.

Still another way to increase the induced voltage is to increase the amount of flux so that for a given speed of the moving conductors more lines of force are cut in a specified unit of time. Increasing the amount of flux means, of course, an increase in the flux density or the lines per square inch cut by the conductors.

Electromagnetic Induction

ALL of the factors involved in the electromagnetic induction of a voltage in a coil actually concern the rate at which the magnetic field is being cut by the conductors of the coil or, to put it another way, the rate at which lines of magnetic flux through a coil change with time. This latter evaluation is the basis of a physical law known as Faraday's Law of Induction, which is stated as follows:

Whenever the number of lines of force linking through a coil is changed, an emf is induced in that coil. And the value of the induced emf is proportional to the rate at which the number of lines of force through the coil is changing.

Experiment indicates that well-defined characteristics exist for the various factors of induction. The numerical value of the induced emf depends upon the density of the magnetic field, the length of the conductor making up the coil and the speed at which the coil conductors cut through the magnetic field. The polarity of the induced emf and the direction of current flow which such emf will produce through an external circuit is determined by the direction or polarity of the magnetic lines of force and the motion of the coil conductors relative to the field. This relationship is generally illustrated in Fleming's Right-Hand Rule, also known as the Generator Right-Hand Rule. This was shown previously by a sketch, but may be stated as follows:

Extend the thumb, forefinger and middle finger of the right hand so that each is at right angles with the other two. Point the forefinger in the direction of the magnetic lines of force, from the North Pole to the South Pole through the air, and point the thumb in the direction in which the conductor is moving (or in the direction in which one side of a rotating coil is moving through the field). The middle finger will then

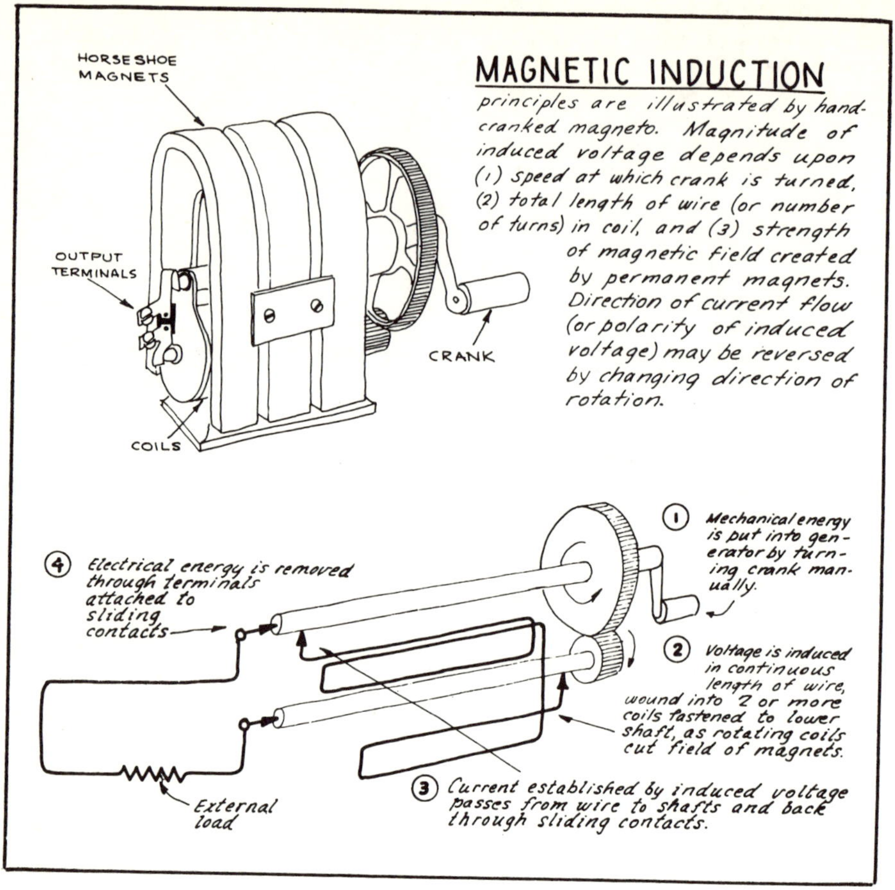

point in the direction of the positive end of the conductor or conductors in which an emf is induced. When the emf so indicated is connected to an external circuit, current will flow in the direction in which the finger is pointing.

This is a very handy and practical rule and should be thoroughly understood for effective application in many problems concerning electromagnetic induction. And although the rule as stated here involves a stationary magnetic field and a moving conductor, just a little care in observing relative motions will permit the use of the rule for studying generator action when the conductor is stationary and the magnetic field moves.

Another law which relates to the matter of electromagnetic induction is Lenz's Law and is an expansion on the observations of Faraday's Law. According to Lenz:

Whenever a current flows as the result of an induced emf, this current sets up a magnetic field about the conductor such that this conductor magnetic field reacts with the magnetic field producing the induced emf to oppose the motion of the conductor. The two magnetic fields react in accordance with the basic precepts

that like poles repel and unlike poles attract. The result is a physical opposition to movement of the conductor through the inducing field, requiring a definite expenditure of external energy in moving the conductor.

The foregoing actions conform to the basic physical law of conservation of energy: To get energy out of anything, we must first put energy into it. In the case of electromagnetic induction, we are getting electric energy out of the conductors moving through the magnetic field. This electric energy is in the form of working current to light a lamp or to do some other work, produced by the induced emf. Then there must have been energy input to the generator and this energy was that required to overcome the opposition to movement of the conductor. In the case of an actual generator, the prime mover—say a steam turbine or a water wheel—must supply the energy to rotate the coils in the magnetic field in opposition to the reaction of the two magnetic fields present. Thus a rotating, mechanical input of energy is converted to electrical energy in the generator, with a certain amount of energy being expended as heat losses in the conversion process.

Self-induced EMF is another very important electrical characteristic which is frequently involved in practical applications of both ac and dc circuits. When any conductor is carrying currents (so far we are considering only direct currents), a magnetic field is set up in concentric rings around the conductor, as we have discussed previously. If the value of current is changed, there is a corresponding change in the magnetic field intensity around the conductor. But any such change in magnetic field density must follow Faraday's Law and therefore produces an induced emf right in the conductor itself.

If the current were reduced, the number of lines linking the conductor would be decreased and there would be an emf induced with a specific polarity. If, however, the current were increased, the number of linking lines would be increased, and there would be an emf induced with opposite polarity to the former case.

In either case, though, the induced emf acts in accordance with Lenz' Law to oppose the change that produced it; i.e., the induced emf is in a direction to oppose the change in current in the conductor in which it

SELF-INDUCTION, *whereby voltage is induced in a conductor itself by the magnet field created by the current flowing through it, takes place only as long as the current is increasing or decreasing. When the current is steady, there is no self-induced voltage.*

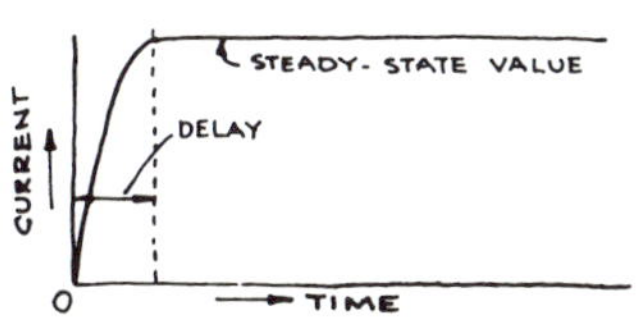

When a circuit switch is closed, current must increase from zero to its steady-state value. Self-induced voltage opposes this increase, causing a slight delay in time required for current to reach this value.

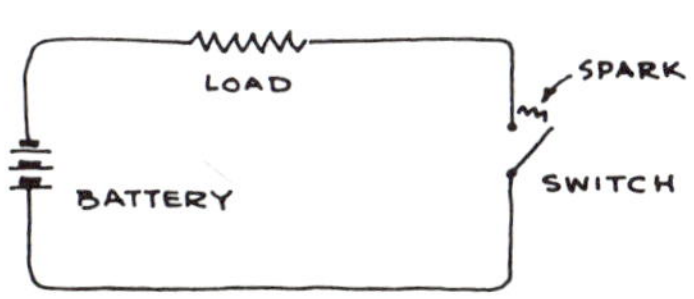

When switch is opened, current must drop back to zero. Self-induced voltage, again opposing the change, attempts to establish a current flow in the opposite direction across air space between switch contacts, creating a spark.

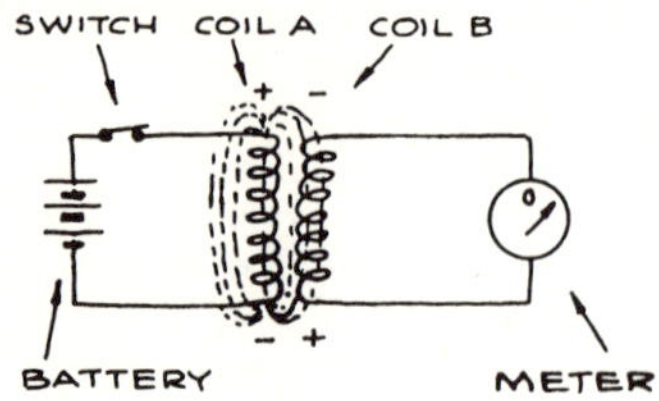

MUTUAL INDUCTION *may be demonstrated by the circuit shown. The coils are not connected electrically and may be wound on the same core. When the switch is closed, current from battery flows through coil A. The resulting magnetic field builds up cutting the conductors of coil B. The needle of a sensitive voltmeter, ammeter, or galvanometer connected across coil B will deflect, indicating that an emf has been induced in coil B. The needle will quickly return to zero once the magnetic field is fully established and constant.*

IN THIS CASE, ELECTROMAGNETIC INDUCTION OCCURS BECAUSE OF THE MOTION OF THE MAGNETIC FIELD: **ALL** CONDUCTORS INVOLVED ARE AT REST.

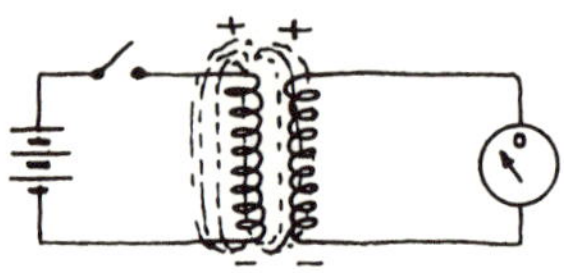

If the switch is now opened, current ceases to flow in coil A and its magnetic field collapses. It again cuts the turns of coil B, inducing an emf in coil B. However, the lines of force now cut the coil in the opposite direction; thus the polarity of the emf in coil B is opposite that established when the switch was closed, and the meter will deflect in the opposite direction.

is induced. Thus, if the current in the original conductor were quickly decreased, the self-induced emf would oppose the decreasing of the current. The emf would add current of its own during the transition from the higher value of current to the lower value. On the other hand, if the current in the original conductor were quickly increased, the self-induced emf would oppose the increase in current flow, retarding the establishment of the higher current level. Of course, in either case, the self-induced emf would be present only for the duration of the transition from one current to the other current. Under steady state conditions of current, there can be no self-induction.

SELF-INDUCED EMF, as discussed previously for a straight conductor, refers to an emf produced in the conductor due to a change in current flowing through the conductor. When a conductor has the property of opposing any change in current flow, the conductor is said to possess **self-inductance** or simply **inductance.**

As explained, a straight conductor possesses a certain amount of inductance by reason of the field around even a single wire. But if a length of wire is wound into a coil, it will possess much greater inductance because of the flux-concentrating effect of current flowing through a coil. As a result, a change of current flowing through a coil will produce much more of a change in the flux linking the circuit. This is due to the fact that each turn of a coil produces a magnetic field which not only acts on itself in inducing a self-induced emf but also acts on adjacent turns of the coil by inducing emf's in them when current flow through the coil changes. The effect of this inductance of a coil is the same as the effect of the inductance of a single conductor; increase in current is opposed by the induced emf and decrease in current produces an emf which tends to keep current flowing. Thus, inductance opposes any change in current flow through a circuit. When current flows at a constant value—without any change up or down in the current with respect to time—the inductance of the circuit has no effect whatsoever on current flow. Under steady-state current conditions, the resistance of the circuit is the only opposition offered to current flow.

Although it has been described so far in terms of varying or fluctuating direct current, inductance as a factor in practical electrical application has most significance in alternating current circuits. This is, of course, due to the fact that in ac circuits the value of current is varying regularly from zero to some positive value and then to some negative value, producing a constantly changing magnetic field about a circuit conductor or about a coil—such as a transformer primary, a fluorescent ballast, a motor winding or any other coil.

In ac circuits, for instance, inductance in circuit conductors—whether service entrance conductors, feeders or branch circuit conductors—contributes to voltage drop from the source of supply to the load. It acts in addition to voltage drop due to resistance, although it is not simply added to the resistive voltage drop. Voltage drop due to inductance is generally significant only in the larger sizes of conductors—such as 1/0 and larger. The full evaluation of inductive voltage drop will be given later under alternating current circuits.

When a switch is used to close the circuit from a dc supply, such as a

TELEPHONE CIRCUIT *illustrates principle of mutual induction. Magnetic field of transmitter coil varies with voice currents; similar currents are induced in receiver coil and are converted to sound in the receiver.*

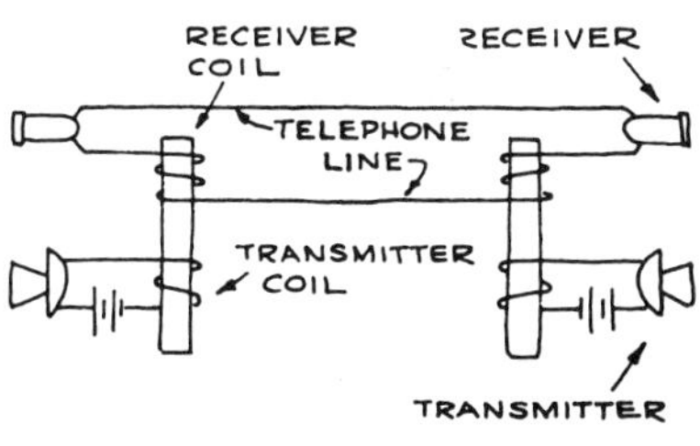

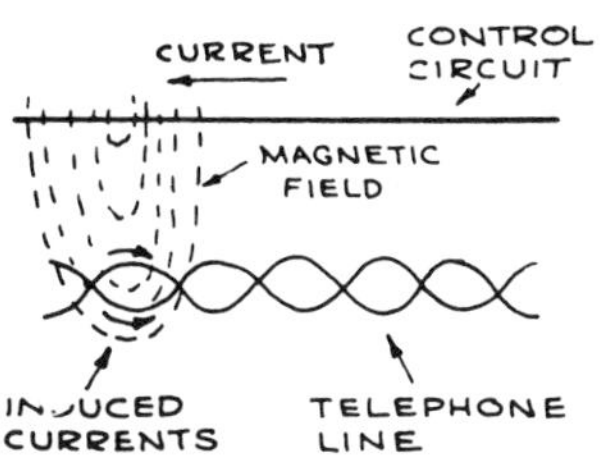

Mutual induction also occurs between parallel conductors. The opening or closing of a switch in a control circuit, for example, could induce a momentary current in an adjacent telephone line, causing a click. By using a twisted pair for the telephone line, equal currents are induced in each of the two telephone conductors, canceling each other out.

battery, to a coil, the self-inductance of the coil delays the build-up of the current from zero to the steady-state value determined by the resistance. The self-induced emf opposes the current build-up. Then, once the direct current has settled down to a steady state value, if the switch is opened, the collapsing magnetic field sets up an induced emf which delays the decay of the current to zero. This tendency to delay decay to zero can often present a problem in the form of an arc across the terminals of the switch. When a highly inductive circuit is opened, the emf of self-induction may be high enough to damage the insulation of the coil, to damage the contacts of the switching device or to present a serious hazard due to arcing. Arc suppression is frequently provided to protect contacts of switches or other devices used to open inductive circuits.

Mutual Induction

So far it has been shown that changing current in a conductor or in a coil results in change in the number of magnetic lines of force linking the conductor or coil and, as a result, sets up an induced emf in the conductor or coil. And such induction is called **self induction.** There is, however, another type of induction by which the varying magnetic field due to changing current in a conductor induces an emf in another conductor placed close enough to the current carrying conductor to be linked by the varying field of that conductor. Or, the varying magnetic field of one coil will induce an emf in another coil placed close enough to the current carrying coil to be linked by the magnetic field. In either case, the induction of an emf in the conductor or coil which is not producing the varying magnetic field is called **mutual induction.**

Mutual induction is actually a form of the previously described phenomenon of generator action. In one case it was shown that an emf could be generated in a coil by rotating the coil in a magnetic field of fixed flux density. In a second case it was shown that the same emf could be generated in the coil by holding the coil fixed in position and rotating the magnets producing the fixed density magnetic field. It is readily seen that in either case there is the necessary relative motion between lines of force and conductors. This meets Faraday's Law: Whenever the number of lines of force linking through a coil is changed, an emf is induced in that coil. But this law can also be satisfied by two coils without relative motion between them, by using the "motion" of a building and collapsing magnetic field instead of the fixed field as used in the two foregoing cases.

If a coil carrying varying current is placed adjacent to another coil which is connected to a voltmeter, the number of lines of force produced by the current-carrying coil and linking with the second coil will vary with current variations in the first coil. The emf produced in the second coil will be indicated on the voltmeter.

The coil to which current is supplied is called the primary coil and the secondary coil. The "coupling" of the primary coil to the secondary coil can be greatly increased by winding both coils on a common iron core to intensify the magnetic field and bring the two coils as physically close together as possible.

Application of the phenomenon of mutual induction is made in circuits having alternating current or pulsating or varying direct current. Probably the most common application is the transformer, in which energy is transferred from the primary coil to the secondary coil—stepping the voltage up, stepping it down or maintaining it at the same value.

Alternating Current

OF ALL the electric energy used today, most of it is of the alternating current (or ac) type.

Although there are many and varied specific applications in which direct current is used today, the characteristics of alternating current electricity suit it to general use for power, light, heat and signals in modern residential, commercial, institutional and industrial buildings. A sound understanding of the nature and applications of alternating current is essential to a practical and effective background in electrical technology.

Basic Concept

A first explanation of alternating current is best given by comparing it to direct current electricity. In a dc circuit, the electron flow through the conductors is always in one direction—or the current flow is direct, from which the name direct current comes. In an alternating current circuit, however, the direction of current flow is constantly reversing on a fixed time basis. These current reversals—or current alternations, to use a more precisely descriptive word—occur in a uniform manner, with a fixed frequency. Consider one cycle of the alternating current flow:

Starting from zero (no electrons moving in the conductor), current begins to flow in one direction through the conductor. Starting slowly, the amount of current flow increases to some maximum value; then, with the same speed, the current decreases to zero. But as soon as the current gets to zero, it does not stop. It immediately makes the same variation in value, from zero to maximum and back to zero, but with the electrons flowing in the opposite direction from that of the first variation. This back-and-forth shuttling of the current is repeated constantly so long as the source of ac voltage is applied to the closed circuit.

From the foregoing description, it is seen that alternating current flows in one direction for a fraction of a second, reverses and flows in the other direction for the same fraction of a second, reverses again, etc. The character of alternating current flow is, of course, determined by the character of the voltage source supplying the circuit. That is, the current is produced by the voltage; and its alternations are the result of the alternations of the voltage and the characteristics of the circuit through which the current flows. Current flow through any conductor will conform to the type of voltage applied. A dc voltage produces dc current. An ac voltage produces an ac current.

Frequency of AC

The most common frequency of alternating current power in use today is 60 cycles per second. Some utilities make available 50-cycle current, and some produce 25-cycle current. Of course, electric power of higher frequencies is widely generated on a local basis for specific applications in commercial and industrial buildings. High-frequency lighting—400 cycles, 840 cycles or even higher frequency—is generally supplied from motor-generator sets or various converters which are themselves supplied by 60-cycle circuits. Various induction and dielectric heating equipment is fed from unit converters at the point of use. And use is made in some industrial plants of portable electric tools operating at frequencies above 60 cycles.

The frequency of an alternating current is usually expressed simply in terms of "cycles". For instance, 60 cycles or 400 cycles. But this must be understood to be a shortened form of "cycles per second". In the case of 60-cycle current, the designation reveals that the current goes through exactly 60 complete cycles of current alternations in each second. Each cycle consists of current variation from zero to maximum to zero in one direction and from zero to maximum to zero in the other direction. As a result, there are 120 reversals of direction of current flow during each second.

Method of Generation

The common type of alternating current power used for general purpose power and light loads is produced by alternating current generators. In an ac generator (or alternator) the rotor usually contains the electro-magnetic field which sets up the magnetic flux lines. When the rotor is rotated by the prime mover driving it (steam turbine, gas turbine, water wheel, steam or combustion engine), the magnetic lines of force are made to cut through the stationary conductors of the coils mounted around the inside of the stator.

As described earlier, this relative motion between lines of force and electric conductors produces an electromotive force in the conductors. Because of the rotating action of the lines of force cutting through the coil conductors and the varying angle at which the lines cut the conductors, the voltage produced in the coils alternates in polarity and varies in intensity during each half cycle. Thus the emf is produced in an alternating voltage. A simplified explanation of the generation of an alternating emf in a single conductor is given in an accompanying sketch, where, for the sake of simplicity, the coils are shown on the rotor of a 2-pole generator and the field poles are represented as fixed magnets.

Variation in the value of generated voltage is shown in the graph. A curve plotted from the instantaneous values during one complete cycle of the 2-pole generator has the same shape as a curve would have which showed the values of the sine of the various angles of rotation. The wave shape is therefore called a "sine

ALTERNATING CURRENT

generation may be understood by considering a single conductor on the armature of a simple 2-pole generator and applying the principles of electromagnetic induction for various positions of the conductor as it rotates in the magnetic field.

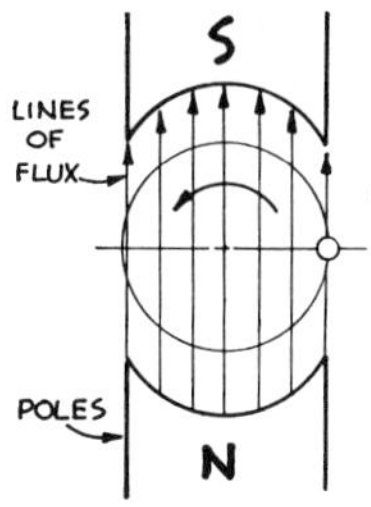

Armature conductor at this point (taken as zero degrees) is moving parallel with magnetic lines. No lines are being cut; no voltage is induced in conductor; no current flows.

As armature rotates, conductor begins to cut flux lines, inducing a voltage and current. Application of right-hand rule shows that current flows into paper, indicated by symbol ⊕

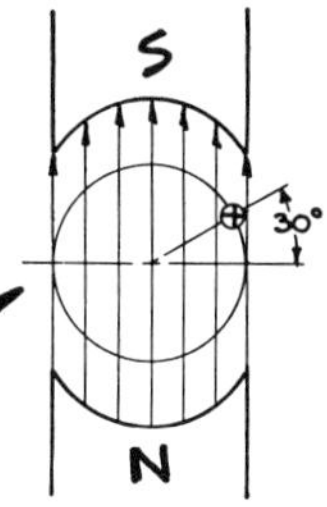

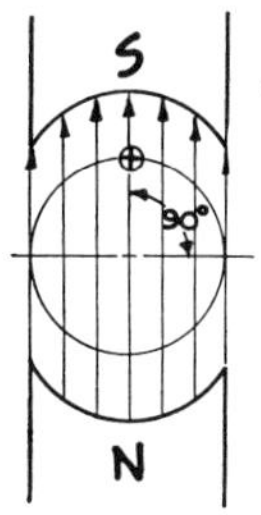

When conductor reaches the 90° point, it is cutting the flux at right angles and hence at a maximum rate. Maximum voltage is induced at this point.

As conductor continues around, the induced voltage decreases; until conductor again moves parallel to flux lines at 180°, and no voltage is induced.

As the degree of rotation passes the 180° point, conductor is again cutting flux, this time from left to right. Right-hand rule shows current now flowing out of paper, indicated by symbol ⊙.

Induced voltage reaches maximum at 270°, then decreases to zero at 360°, at which time a new cycle begins.

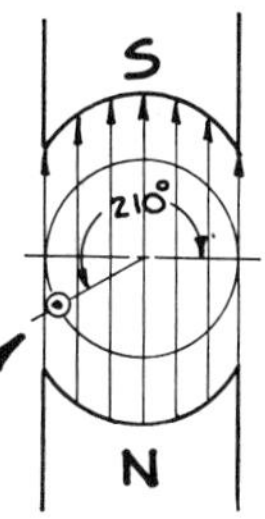

MAGNITUDE OF VOLTAGE (OR CURRENT) AT VARIOUS POINTS OF ROTATION MAY BE PLOTTED AS SHOWN. CONNECTING THE PLOTTED POINTS GIVES THE FAMILIAR SINE WAVE ASSOCIATED WITH ALTERNATING CURRENT.

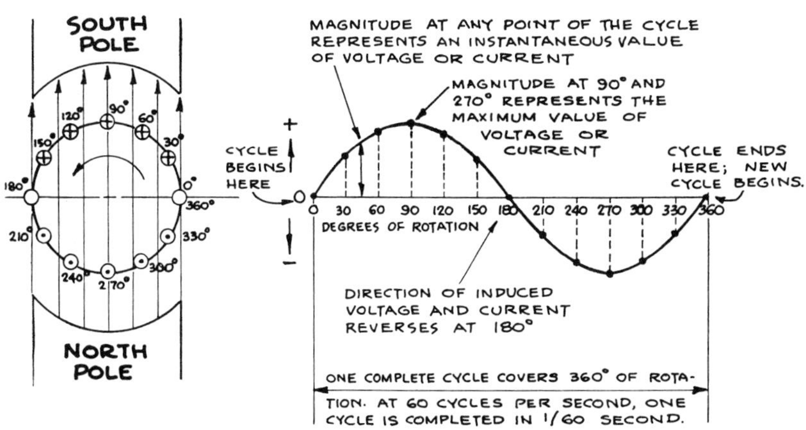

wave". All references to alternating current for power and light electrical systems are based on the assumption that voltages and currents are sine waves. All references to alternating half cycles have the same shape but successive ones differ in polarity. Those above the line are commonly referred to as positive alternations and those below the line as negative alternations.

During the alternations of an alternating voltage or current, the value at any given time is called the "instantaneous" value; and the greatest value reached in each half cycle is called the "maximum" value.

AC Fundamentals

BECAUSE of the comparatively complex nature of alternating current circuits, there are many specific characteristics of ac based on the varying values of current during each cycle. The important characteristics which must be clearly understood for effective application of ac are as follows:

• **Electrical Degrees** — As shown previously, each cycle of current alternation takes place in a fixed period of time, depending upon the frequency of the ac. For instance, each cycle of a 60-cycle ac occurs in 1/60 of a second. A half cycle at this frequency occurs in 1/120 of a second. At any other frequency, the time interval for a cycle or part of a cycle will, of course, be different.

But there is a way of referring to a cycle or part of a cycle which is not related to frequency, but which greatly facilitates other analysis of ac, such as phase relationships. This is the use of electrical degrees. One complete cycle is 360 degrees; one-half cycle, 180 degrees; etc. Each cycle starts at 0 degrees. The positive alternation or half-cycle occurs as the vector generating the sine wave moves from 0 degrees to 180 degrees. The negative alternation or half-cycle occurs from 180 degrees to 360 degrees. It is important to note that the units of degrees when applied to a current or voltage wave are actually representing units of time, inasmuch as a cycle is a current variation with time.

• **Effective Value of Alternating Current**—As described previously, the value of current in an alternating current wave is changing constantly—i.e., the instantaneous value is changing from zero to maximum to zero, with alternate directions of current flow. The instantaneous value of current at any interval of electrical degrees in a cycle is equal to the maximum value of current for the wave multiplied by the sine of the vector angle, which corresponds to the value of electrical degrees for the interval at which the current is taken. Instantaneous value of voltage is obtained in the same manner. It is the product of the maximum value of voltage times the vector angle. But there is little practical use of the instantaneous values of current and voltage in electrical circuits. The more significant values are the effective values of current and voltage.

Because an alternating current is constantly varying in instantaneous value and regularly alternating in direction, the only way to evaluate the current is to determine its effective value. This can be done by determining the heating effect of an alternating current and giving the ac current the

MAXIMUM AND EFFECTIVE AC VOLTAGE AND CURRENT

The common ac generator produces a continuously variable voltage and current the instantaneous values of which, when plotted against time or degrees of rotation, form a sine wave.

In the diagram, the angle θ represents electrical degrees of rotation of a point on a conductor moving through a magnetic field about the center axis of the armature.

The MAXIMUM VALUE of the generated voltage or current (E_m or I_m) is proportional to the hypotenuse "c" of the triangle, which is the radius of the circle.

The INSTANTANEOUS VALUE of the voltage or current (E_i or I_i) is proportional to the side "a" opposite angle θ.

Note that $a = 0$ at 0°, 180°, and 360°.
Also $a = c$ at 90°; $a = -c$ at 270°.

Thus, $E_i = 0$ at 0°, 180°, and 360°
and $E_i = E_m$ at 90°; $E_i = -E_m$ at 270°.

At all other points, E_i will be some value between 0 and E_m during the positive half of the cycle and between 0 and $-E_m$ during the negative half.

We can find the value of E_i at any point of the cycle using a table of sine values and the formula

$$E_i = E_m \sin\theta$$

Similarly,

$$I_i = I_m \sin\theta$$

Through higher mathematics, it is found that the ac EFFECTIVE CURRENT – that current producing the same heating effect as a dc current of equal magnitude – may be found as follows:

$$I_e = 0.707 I_m$$

The same relationship applies to ac voltages:

$$E_e = 0.707 E_m$$

EXAMPLE 1: The sine wave of an alternating current shows a maximum value of 100 amperes. What value of dc current will produce the same heating effect?

$$I_e = 0.707 I_m$$
$$= 0.707 \times 100$$
$$= 70.7 \text{ amperes.}$$

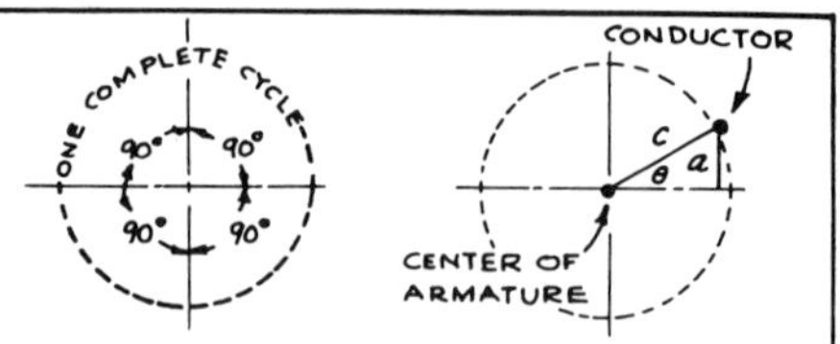

A complete circle has 360°; thus 1 complete revolution of a point on a conductor on an armature covers 360°. At 60 cps, 1 cycle takes 1/60 sec., 1/2 cycle takes 1/120 sec., etc.

From trigonometry, the sine (abbreviated "sin") of an angle is defined as the side opposite (a) divided by the hypotenuse (c).

Thus, $\sin\theta = \frac{a}{c}$

Also, $c = \frac{a}{\sin\theta}$

and $a = c \sin\theta$

EXAMPLE 2: The maximum instantaneous voltage produced by a generator is 170 volts. Find instantaneous voltage values at 30° intervals over the complete cycle and draw the sine wave representing the voltage. What is the effective voltage?

TABLE OF SINES	
θ	SIN θ
0°	0
30°	0.5
60°	0.866
90°	1.0
120°	0.866
150°	0.5
180°	0
210°	−0.5
240°	−0.866
270°	−1.0
300°	−0.866
330°	−0.5
360°	0

At 30°, $E_i = E_m \sin\theta$
$= 170 \sin 30°$
$= 170 \times 0.5$
$= 85$ volts.

Continuing this procedure for all angles shown in the table gives the following:

θ	E_i	θ	E_i	θ	E_i
0°	0	120°	147	270°	−170
30°	85	150°	85	300°	−147
60°	147	180°	0	330°	−85
90°	170	210°	−85	360°	0
		240°	−147		

Plotting this data gives the sine wave:

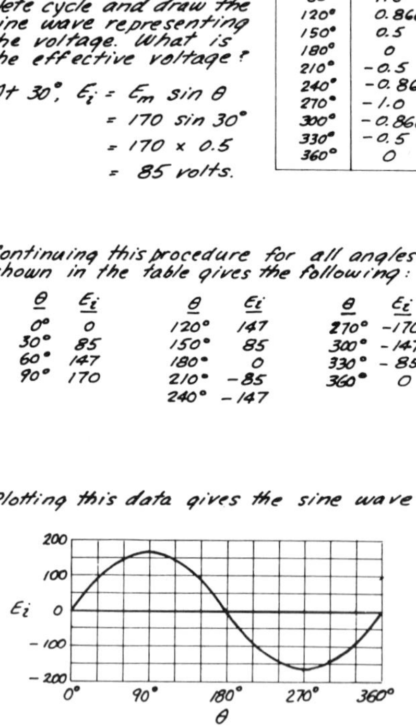

The effective voltage in this case will be
$E_e = 0.707 \times 170 = 120$ volts.

This is the rating which would appear on the machine nameplate.

same ampere designation as a direct current which will produce the same heating effect.

As discussed in the study of direct currents, the heating effect of direct current is an expenditure of power and is equal to the square of the current times the resistance of the current path, i.e., $P=I^2R$. From this comparison, it can be stated that an alternating current has an effective value of one ampere when it will produce the same heat in a given resistance as one ampere of direct current.

The value of effective current is given in amperes. Another common description of this effective value is the "rms" value. The letters "rms" stand for "root-mean-square." This expression derives from the comparison of ac heating effect to dc heating effect.

A dc current produces heat at a rate equal to I^2R. The heating effect of an ac current half cycle is being produced at any instant at a rate equal to i^2R, where "i" is the instantaneous value of current at each instant. The heating effect of each half cycle depends upon the average of the squares of the instantaneous current values for the half cycle. By taking the square root of the mean or average square of the instantaneous current values, the effective or rms value can be determined. This value then represents the heating effect, or rate of heat production, during each half cycle, regardless of the direction of current flow. And so long as the current alternations of the same value follow one after the other, the effective or rms value of the current is maintained at a steady level.

The effective or rms value of a sine wave alternating current is equal to 0.707 times the maximum, or peak, value of the current wave.

It is important to remember that when an alternating current or ac voltage is given a value it is always the effective value unless specific designation is given to the contrary. Equipment nameplates, system designations, voltage and current ratings of circuits and equipment, readings of test instruments and just about all practical references to voltage and current are based on effective values.

• **Effective Value of AC EMF**—The alternating currents discussed so far are caused to flow by voltage, or EMF, which is the cause of the alternations in current flow. The same relationships exist, therefore, among instantaneous, maximum and effective or rms values of voltage, e.g., rms value of a sine wave.

Ohm's Law for AC

Application of an ac rms voltage to a circuit containing resistance will produce an rms value of current flow in accordance with Ohm's Law: $I=E/R$. Because rms voltage and current values correspond in their nominal values to direct current values which will produce the same heating effect, the application of Ohm's Law can be made to ac circuits in the same way it is made to dc circuits. For instance, 120 volts rms ac applied to a circuit of 12 ohms resistance will produce a current of 10 rms amperes, just as 120 volts dc would produce a 10-amp dc current in the same resistance.

Power in any ac circuit supplying a resistive load can be computed from the same formula used for dc power, i.e., $P=I^2R$, in which "P" is the power in watts, "I" is the rms value of current and "R" is the resistance through which the current is flowing. Variations on this basic power formula are, as explained in the study of dc, the following: $P=E^2/R$ and $P=EI$. As an example, what is the current drawn by a 1,000-watt load of incandescent lamps connected on a 120-volt ac branch circuit? Using the formula $P=EI$ and transposing, $I=P/E$, or $I = 1000/120 = 8.3$ amperes.

AC Circuit Analysis

IN ALTERNATING current circuits, some very important considerations arise from the waveform character of the current and the voltage. A basic consideration is that of the phase relation between the current and the voltage waveforms.

When an alternating current voltage is applied to a circuit containing resistance only, the current wave is produced by the voltage wave. In such a circuit, the current and voltage pass through their zero values and increase to their maximum values in the same direction at the same time. In that case, the current is said to be "in phase" with the voltage. However, in circuits which contain opposition to current flow other than resistance (such as magnetic coils or capacitor effects—to be discussed later), the current and voltage waves do not go through their zero and maximum values at the same time. Instead, there is a fixed time interval between zero and maximum values of current and zero and maximum values of voltage. In this latter case, the current is said to be "out of phase" with the voltage.

There are two possible "out of phase" relations between voltage and current:

1. The current may pass through its zero value and increase to a maximum at some time later than the voltage has passed through its zero and increased to its maximum. In such a case, the current is said to "lag" the voltage.

2. The current may pass through its zero value and increase to a maximum at some time earlier than the voltage passes through its zero and increases to its maximum. In this case, the current is said to "lead" the voltage.

Depending upon the particular conditions of a circuit, a current may lead or lag the voltage by any value up to ¼ cycle or 90°. Using electrical degrees as a measure of time (as described previously), the difference in time or the "phase difference" is commonly expressed in electrical degrees and is called the "phase angle" of the circuit. The Greek letter "theta"—Θ—is the usual symbol for phase angle.

Phase difference, as described above, is an out-of-phase relation between a voltage and the current it produces. However, there may also be circuits in which two or more voltages have phase differences between them, or two or more currents exist with phase differences. Such conditions will be found in polyphase alternating current circuits — three-phase circuits, for example.

Load Characteristics

When a dc voltage is applied to a circuit, the amount of current which flows is determined wholly by the resistance in the circuit. However, when an ac voltage is applied to a circuit, the amount of current which flows is determined not only by the resistance of the circuit but also by the amount of inductance and capacitance. Resistance opposes current flow in ac circuits in much the same way as it opposes current flow in dc circuits. This was shown previously in the case of Ohm's Law for ac. But inductance and capacitance are circuit characteristics which affect current only in ac circuits.

When an ac circuit contains only resistance, the current will be in phase with the voltage, i.e., the phase angle will be zero. And the circuit will conform to Ohm's Law.

Inductance—As discussed in a previous section on "Electromagnetic Induction," inductance is the property of a circuit or circuit element, such as a coil or other electromagnetic device, which opposes any change in the current flowing in the circuit. This opposition arises from self-induced emf due to varying flux resulting from the alternating character of the current. Any part of a circuit which produces flux has inductance. This includes all forms of coils wound on magnetic cores—motors, ballasts, transformers, solenoids, etc.

CURRENT FLOW THROUGH A RESISTANCE *encounters no opposing emf; hence circuit voltage and current rise and fall together, or "in phase."*

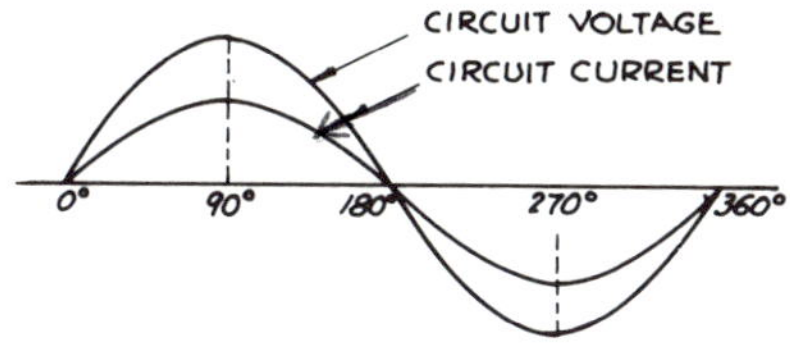

CURRENT FLOW THROUGH AN INDUCTANCE, *however, sets up a magnetic field about the coil which induces an opposing emf in the coil and causes the circuit current to lag behind the circuit impressed voltage.*

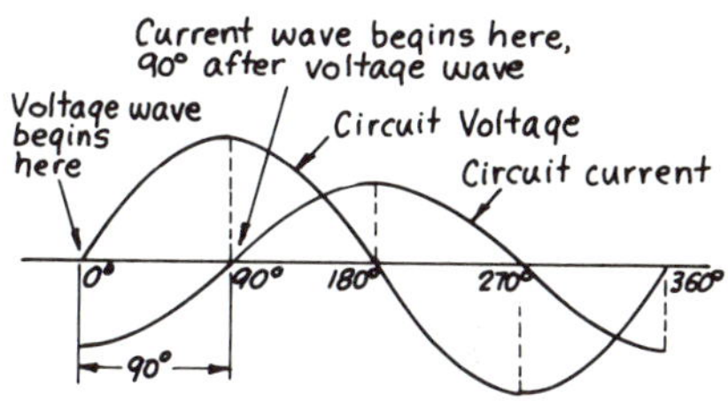

Inductance is measured in basic units known as "henrys". The symbol for a henry is "L" and a circuit or device is said to have an inductance of one henry if, when the current is changing at the rate of one ampere per second, the average induced counter emf is one volt.

Inductive Reactance—In an ac circuit containing inductance, an emf of self-induction is being induced con-

INDUCTIVE REACTANCE

is the opposition, expressed in ohms, presented to the flow of ac current due to the inductance in the circuit:

$$X_L = 2\pi fL$$

X_L = *inductive reactance (ohms)*
π = *3.14*
f = *frequency (cycles per second)*
L = *inductance (henrys)*

$$I = \frac{E}{X_L}$$

I = *effective current (amperes)*
E = *effective voltage (volts)*

EXAMPLE: What effective current will flow through the coil in the following circuit? (Resistance of coil is negligible.)
Note: 1 milli-henry (mh) = $\frac{1}{1000}$ henry.

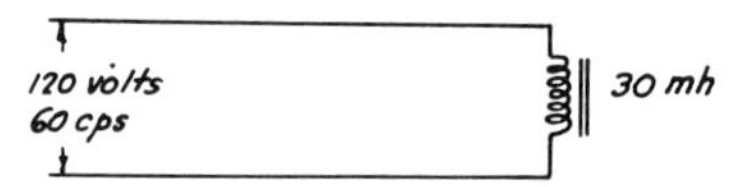

SOLUTION:

$X_L = 2\pi fL = 2 \times 3.14 \times 60 \times .03$
$= 11.3$ ohms

$I = \frac{E}{X_L} = \frac{120}{11.3} = 10.6$ amperes

tinuously because the instantaneous value of current is alternating in value continuously. Because this emf of self-induction is constantly opposing the alternations in current, it is, in effect, opposing the ac current and its effect can be measured in ohms. The opposition that inductance presents to the flow of alternating current is known as "inductive reactance," designated by the symbol "X_L."

In any circuit, the amount of inductive reactance depends upon the value of inductance and the frequency of the alternating current. Thus, the value of inductive reactance increases with increase in inductance and with increase in frequency of current alternations. The value of inductive reactance in ohms may be found from the following formula:

$$X_L = 2\pi fL$$

where

$\pi = 3.14$
f = frequency of current in cps
L = value of inductance in henrys

In a circuit containing only inductance (resistance is negligible), the only opposition to current flow is the emf of self-induction. As a result, the voltage applied to the circuit will be equal and opposite to the emf of self-induction. The value of self-induced emf will be maximum when change in current (and, hence, change in flux) is maximum. This will be when the current wave is going through its zero point. The value of self-induced emf will be zero when the rate of current (and, again flux) change is zero. This latter condition occurs at the maximum value points in the current wave. From all of these conditions, it is found that the current lags the applied voltage by 90°in a purely inductive circuit. And if there is any resistance in a circuit containing inductance (which is usually the case) the angle of current lag will be less than 90°, depending upon the amount of resistance and inductance.

WHEN a voltage is applied to a circuit containing only inductance (i.e., resistance is negligible), the magnitude of current flow is determined from the value of the voltage and the value of inductive reactance in accordance with the following:

$$I = \frac{E}{X_L}$$

where, I = current in amps,
E = applied voltage
X_L = inductive reactance

For example, if 120 volts, 60-cycle ac, is applied to a circuit containing 0.02 henry of inductance, the inductive reactance is equal to 2 x 3.14 x 60 x 0.02 from the formula for inductive reactance, $X_L = 2\pi fL$. This works out to 7.54 ohms. Then, according to $I = E/X_L$, the current flowing in the circuit will be equal to 120/7.54 or 15.9 amps.

Resistance and Inductance

As stated previously, circuits containing inductance and only a negligible amount of resistance are extremely rare. And it should be noted that practical electrical circuits which contain no resistance at all are impossible to achieve. Any conductor of electric current which provides the inductance in an ac circuit has some resistance due to frictional effects of electron flow through the molecular adhesion of the conductor material itself. So consideration must next be given to circuits combining resistance and inductance in series.

A very wide range of electrical appliances and devices present a combination load of resistance and inductance to current flow through them. A motor, for instance, has windings which have a certain amount same can be said of fluorescent lighting ballasts, transformers, solenoid coils and electric generators. The ap- of resistance and also have an appreciable amount of inductance. The preciable amount of inductance in such devices arises from the fact that they all contain wires wound in the form of coils and have magnetic cores which intensify the flux produced by current flow and therefore intensify the inductance. Of course, any conductor carrying current, either ac or dc current, is subject to the effects produced by variations in the flux field set up around the conductor. But for 60-cycle ac circuits, inductance is so small as to be negligible in resistive load devices — incandescent lamps, resistance heating elements — and in the vast majority of circuit conductors. In larger sizes of conductors, such as sizes No. 1/0 and larger, the ratio of inductance to resistance becomes significant due to increased spacing between centers of conductors and the effect of inductive reactance must be taken into consideration in computing voltage drop in such circuits.

In discussing resistance in ac circuits, it is important to note that electric conductors offer greater resistance to the flow of alternating current than they do to the flow of direct current. This difference in resistance of a given wire to ac and dc arises from two sources.

One source of the increase in resistance is called "skin effect," deriving the name from the fact that alternating current tends to flow along the outside, or "skin," of a conductor. This is due to inductive action which forces the current to the outside of the wire. But it must be made clear that although skin effect and inductive reactance are both produced by the fluctuating magnetic field, they are completely separate and distinct effects. Skin effect is resistive opposition and will not shift the phase relation between current and voltage, Inductive reactance, as explained, is a back-emf type of opposition to current flow and does shift the phase between current and voltage.

A second source of increase in resistance for ac is called "proximity effect." This effect refers to the fact that the varying magnetic field set up around a conductor carrying ac will produce losses due to currents induced in adjacent conductors, due to hysteresis losses in any magnetic material (such as steel conduit) around the conductor, due to currents induced

IMPEDANCE IN INDUCTIVE CIRCUITS

Current through a resistance is in phase with the voltage; current through an inductance lags the voltage by 90°. The resistance (R) and the inductive reactance (X_L) which provide opposition to these currents may also be thought of as being 90° apart. The net opposition to the current, impedance (Z), will then be represented by the hypotenuse of the triangle thus formed. Since the square of the hypotenuse of a triangle is equal to the sum of the squares of the other two sides,

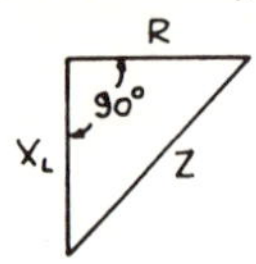

$$Z^2 = R^2 + X_L^2.$$

Therefore, in a series circuit containing both resistance and inductance,

$$Z = \sqrt{R^2 + X_L^2}$$

Impedance, like resistance and reactance, is expressed in ohms. It represents the apparent resistance of a circuit to alternating current, so that

$$\text{EFFECTIVE CURRENT} = \frac{\text{EFFECTIVE VOLTAGE}}{\text{IMPEDANCE}}$$

or

$$I_{EFF} \text{ (AMPERES)} = \frac{E_{EFF} \text{ (VOLTS)}}{Z \text{ (OHMS)}}$$

AC RESISTANCE

MULTIPLYING FACTORS FOR CONVERTING DC RESISTANCE TO 60-CYCLE AC RESISTANCE (Copper conductors in metallic-sheathed cables or in metallic raceways)

SIZE	FACTOR
3 AWG	1.00
2	1.01
1	1.01
0	1.02
00	1.03
000	1.04
0000	1.05
250,000 CM	1.06
500,000	1.13
1,000,000	1.30
2,000,000	1.82

The above data, extracted from the National Electrical Code, indicates the extent to which dc and ac resistances differ. The difference is less pronounced for non-metallic sheathed cables, cables in air or in non-metallic conduit, and for aluminum cables.

in metallic raceways, either magnetic or non-magnetic. Because these power losses must be supplied by the conductors producing them, they increase the effective resistance of the circuit over what it would be for direct current, which does not produce the above losses.

Both skin effect and proximity effect exist to some extent in all conductors carrying alternating current. But this increase in resistance is so small for most circuits that it need not be considered. That is, the resistance of conductors to ac can be taken to be the same as the dc resistance for all conductors smaller than 500 MCM.

For conductors in which skin effect must be included in evaluating resistance, the ac resistance can be determined by multiplying the dc resistance by the factor given in handbook tables on dc to ac resistance conversions. One such table is given in Chapter 9 at the end of the National Electrical Code.

When a voltage is applied to a circuit containing resistance in series with inductance, calculation of the current flow follows a fixed procedure: First the value of inductance in henrys is taken and converted to an ohms value of inductive reactance by $X_L = 2\pi fL$. Then the ac value of the resistance must be combined with the inductive reactance value in a way which accounts for the differing phase relations. The net effect of the resistance and inductive reactance acting in series is to produce an opposi-

MCM = Thousands circular mils

EXAMPLE:

For the circuit shown, find

(A) THE INDUCTIVE REACTANCE
(B) THE IMPEDANCE
(C) THE EFFECTIVE CURRENT
(D) THE MAXIMUM CURRENT

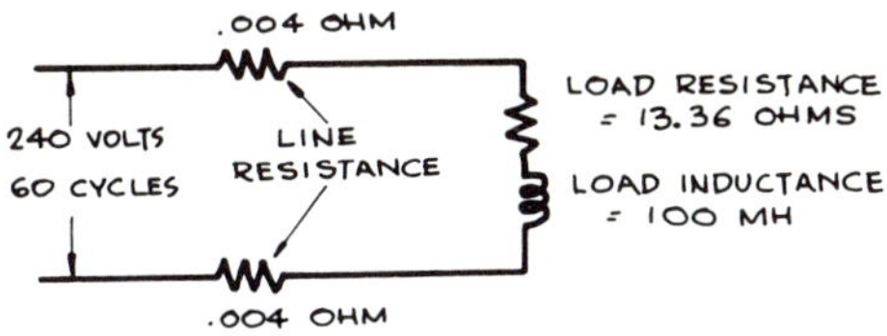

SOLUTION:

(A) INDUCTIVE REACTANCE $X_L = 2\pi f L$

$X_L = 2 \times 3.1416 \times 60 \times 0.1$

$= \underline{37.7 \text{ OHMS}}$

(B) IMPEDANCE $Z = \sqrt{R^2 + X_L^2}$

RESISTANCE $= 13.36 + .004 + .004$

$= 13.368$ OHMS

$Z = \sqrt{(13.368)^2 + (37.7)^2}$

$= \sqrt{178.7 + 1421.3}$

$= \sqrt{1600} = \underline{40 \text{ OHMS}}$

(C) EFFECTIVE CURRENT $= \dfrac{E_{EFF}}{Z}$

$I_{EFF} = \dfrac{240 \text{ VOLTS}}{40 \text{ OHMS}} = \underline{6 \text{ AMPERES}}$

(D) EFFECTIVE CURRENT $= 0.707 \times$ MAX. CURRENT

$I_{MAX} = \dfrac{I_{EFF}}{0.707}$

$= \dfrac{6}{0.707} = \underline{8.49 \text{ AMPERES}}$

tion to current flow called "impedance," designated by the symbol "Z." Impedance is found from the following:

$$Z = \sqrt{R^2 + X_L^2}$$

where, Z = impedance in ohms,
R = resistance in ohms
X = inductive reactance in ohms

Then the circuit current can be found from the formula —

$$I = \frac{E}{Z}$$

This formula is the ac form of Ohm's Law and applies to all ac circuits. An example of this calculation is shown in the accompanying illustration.

Circuit Impedances

AS DISCUSSED previously, current flowing through a resistance is in phase with the voltage causing the current flow; but current flowing through an inductance is 90 deg out of phase with the voltage wave producing the current flow, with the current lagging the voltage.

When a resistance and inductance are connected in series across a voltage source, a single current flows with an in-phase relation to the voltage drop across the resistance and a 90° lagging phase relation to the voltage drop across the inductance. The voltage drops—the one across the resistance and the one across the inductance—are therefore 90 deg out of phase with each other. As a result, the simple mathematical sum of the two voltage drops is not equal to the source voltage impressed on the circuit.

The source voltage across the above described circuit is equal to the vector sum of the voltage drop across the resistance and the voltage across the inductance in accordance with the following formula:

$E_S = \sqrt{E_R^2 + E_L^2}$, where

E_S = voltage impressed on circuit.
E_R = voltage drop across resistance, and
E_L = voltage drop across the inductance.

An example of the voltage relationships in a circuit containing resistance and inductance in series is given in the accompanying illustration. It should be noted that the impedance represented by the resistance and the inductance in series is found from a similar formula in which consideration is given to the phase difference between the voltage drops.

Other variations on the basic formulas for circuits containing resistance and inductance in series are as follows:

$$E_S = \sqrt{E_R^2 + E_L^2} = \sqrt{(IR)^2 + (IX_L)^2}$$
$$= \sqrt{I^2(R^2 + X_L^2)} = I\sqrt{R^2 + X_L^2}$$

$$I = E/\sqrt{R^2+X_L^2} = E/Z$$

Capacitance

In addition to inductance, there is another characteristic of importance in ac circuits. This is "capacitance," which has a reactive load effect somewhat similar to inductance but with a different phase relationship between current and voltage.

Capacitance is an electrical characteristic based on the phenomenon of electrostatics. So far we have discussed electric current flow, which is electricity in motion. But there is another form of electricity which is at rest and does not involve a current flow. This is static electricity.

If a source of dc voltage is connected across two conducting plates which are placed close together but are separated by air or other insulating material, the plate connected to the plus terminal will have a positive charge (a deficiency of electrons) and the plate connected to the minus terminal will have a negative charge (a surplus of electrons, the ones drawn from the positive plate by the voltage source). Thus an electrostatic charge exists across the plates. The slight movement of electrons which set up the electrostatic charge indicates the ability of the plates to store an electric charge. If the source of voltage is removed, the plates will hold their charge until the displaced electrons can find a way to leak back to the positively charged plate. Any two conductors separated by a dielectric material — an insulating material which permits passage of electrostatic lines of force—is a capacitor and has capacitance or ability to store electric charges.

Capacitors are devices used in electric circuits and equipment to obtain certain characteristics based on their ability to hold electric charges. Typical capacitor units are made for power factor correction on circuits and for many uses in electronics. The most common dielectric materials used are air, impregnated paper, mica, ceramic and electrolytic film. But in addition to devices intended to provide capacitance, there are sources of capacitance which are accidental. These latter include: capacitance between insulated conductors in contact

R-L CIRCUIT VOLTAGE

In the simple dc series circuit shown, use of Ohm's Law gives the voltage drops across the resistances:

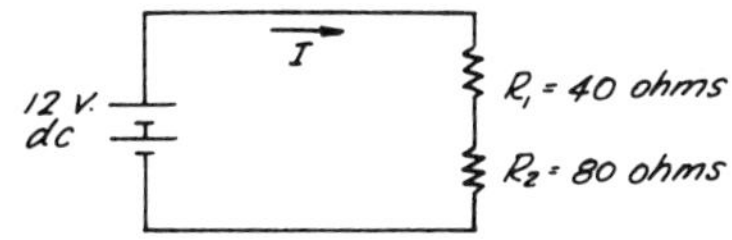

TOTAL RESISTANCE = 40 + 80 = 120 OHMS

CURRENT $I = \frac{12 \text{ VOLTS}}{120 \text{ OHMS}} = 0.1$ AMP

E_1 = 40 OHMS × 0.1 AMP = 4 VOLTS

E_2 = 80 OHMS × 0.1 AMP = 8 VOLTS

$E_1 + E_2$ = 12 VOLTS

Thus in a dc series circuit the sum of the voltage drops add up to the source voltage.

However, in an ac circuit with resistance and inductance in series, the voltage drops do not add up to the line voltage:

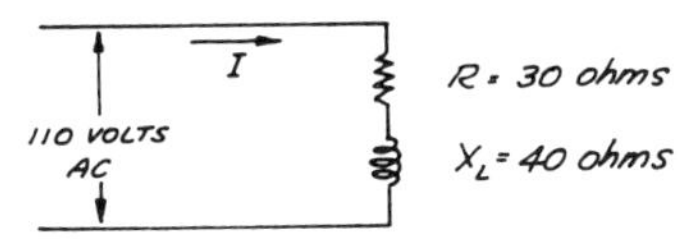

IMPEDANCE $Z = \sqrt{R^2 + X_L^2} = \sqrt{30^2 + 40^2}$

$= \sqrt{900 + 1600} = \sqrt{2500} = 50$ OHMS

CURRENT $I = \frac{E}{Z} = \frac{110}{50} = 2.2$ AMPS

VOLTAGE DROP ACROSS R = IR

= 2.2 × 30 = 66 VOLTS

VOLTAGE DROP ACROSS L = IX_L

= 2.2 × 40 = 88 VOLTS

Note that 66 + 88 = 154 volts, much higher than the circuit's 110 volts. These voltages must be added "vectorially" to get the circuit voltage, using the formula $E = \sqrt{E_R^2 + E_L^2}$.

$E = \sqrt{66^2 + 88^2} = \sqrt{4356 + 7744}$

$= \sqrt{12{,}100} = 110$ VOLTS

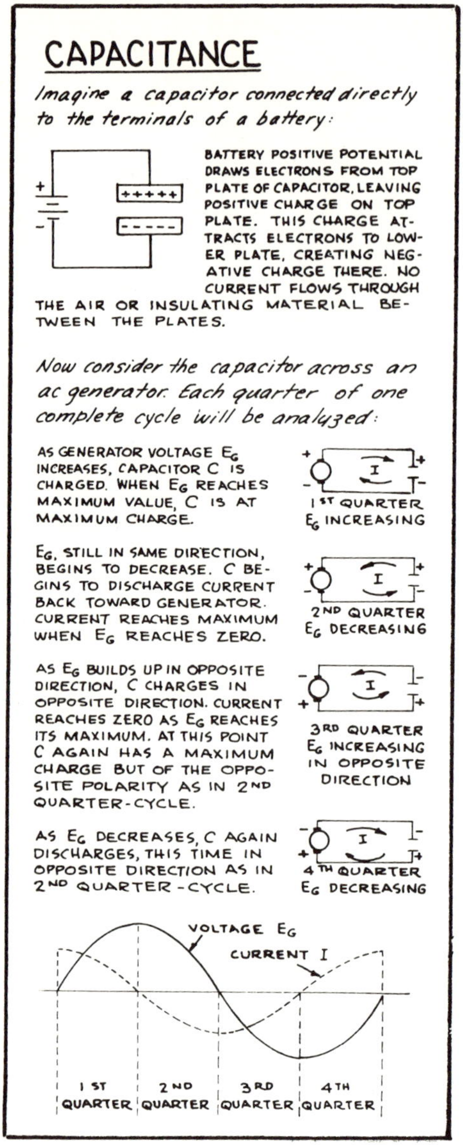

with each other, with the insulation acting as the dielectric; capacitance in the turns of transformer winding; capacitance between any closely placed conducting parts with potential difference between them.

In ac circuits, capacitance—from either capacitors or from stray accidental effects—is an important consideration due to its reaction to alternating current. When a dc voltage is impressed across a capacitor, there is a slight initial rush of electrons to set up the charged condition. Then there is no more electron movement, and the circuit is effectively an open circuit due to the insulator separating the capacitor plates. But when an alternating current voltage is impressed across a capacitor, the back and forth motion of electron flow alternately charges the capacitor with one polarity and then with the opposite polarity. In effect, the capacitor permits alternating current to flow but does act as an opposing effect which reduces the current value from what it would be if the capacitor were not in the circuit.

As shown in a sketch, when an ac voltage wave starts from zero, current starts flowing into the capacitor, storing up the electrons. But the value of current will be decreasing continually from its initial value during the rise of the voltage wave, because as the dielectric stores more and more electrons, it can take fewer and fewer additional electrons. When the voltage wave is at its top, where there is no longer an increase in the potential applied, the flow of electrons into the dielectric ceases. Then as the voltage wave decreases to zero, the electrons stored in the dielectric are caused by the charge on the capacitor to start flowing back, out of the capacitor, in the direction opposite to that in which they entered. And the rate of this outward flow increases from zero to a maximum value which occurs when the impressed circuit voltage has decreased to zero. Electrons then start flowing into the dielectric from the other side of the capacitor. The same process is then repeated with the opposite polarity of the voltage wave and opposite polarity in the charging of the capacitor. A close look at this diagram reveals that the ac current "through" the capacitor "leads" the voltage impressed by 90 deg. This then is a reactive effect and is directly opposite to that produced by inductance.

Capacitive Reactance

A CLEAR qualitative and quantitative understanding of the phenomenon of capacitance can be obtained by considering the action which takes place when a dc voltage source, such as a battery, is connected across the plates of a capacitor:

As described, a charge is produced on the capacitor equal to the voltage of the battery—with a positive charge on the capacitor plate connected to the positive battery terminal and a negative charge on the plate connected to the negative battery terminal. This charge was produced by a minute, momentary flow of electrons when the dc source was connected to the capacitor. And the current flow stopped quickly, leaving the charge on the capacitor. And if the battery is disconnected from the capacitor, the charge would remain on the capacitor provided there was no current path from one plate of the capacitor to the other.

If the charged plates of the capacitor are then shorted—connected together by a piece of wire—the same minute, momentary current would flow as the surplus electrons on the negative plate were attracted by the positively charged ions of the positive plate of the capacitor. The plus and minus charges would be neutralized and the capacitor would be discharged.

In a circuit with a battery of given voltage, the use of a capacitor of a given size will produce a certain amount of current flow when the circuit is closed. Use of a different size capacitor in the same circuit will produce a different momentary current. Thus, the size of a capacitor indicates the effect it will have as a device for storing electrons.

Every capacitor is rated according to the amount of electrons that are stored in it when a source of one volt is connected across it. This ability to store electrons is called "capacitance" and is expressed in units called "farads." A capacitor has a capacitance of 1 farad when an applied source of one volt causes the capacitor to take a charge of 1 coulomb. Because the farad is too large a unit of rating for practical usage, the term "microfarad" (one millionth of a farad) is more commonly used. The symbol

CAPACITANCE

Remember that current does not flow __through__ a capacitor. The constantly reversing voltage causes a "back-and-forth" movement of electrons to alternately charge the capacitor first in one direction, then the other.

Conventional current is normally represented as moving __away__ from the positive terminal of the source, opposite to the actual direction of electron flow:

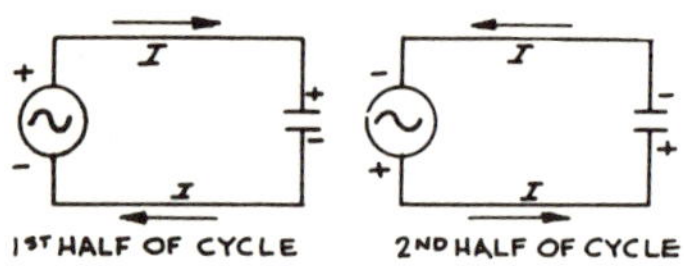

I = effective current

__CAPACITANCE = 1 FARAD__ *when 1 volt produces a charge of 1 coulomb on the capacitor.*

__A COULOMB IS A QUANTITY__ *of electricity, equal to approximately 6 million million million electrons.*

__1 FARAD = 1,000,000 MICROFARADS,__ *abbreviated MFD, mfd, or μfd.*

CAPACITORS IN SERIES

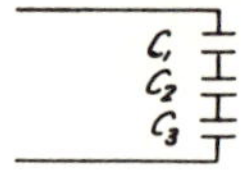

Total capacitance of circuit = C

$$\frac{1}{C} = \frac{1}{C_1} + \frac{1}{C_2} + \frac{1}{C_3}$$

EXAMPLE: FIND C IF C_1 = 20 MFD, C_2 = 30 MFD, C_3 = 40 MFD.

SOLUTION:

$$\frac{1}{C} = \frac{1}{C_1} + \frac{1}{C_2} + \frac{1}{C_3} = \frac{1}{20} + \frac{1}{30} + \frac{1}{40}$$

$$\frac{1}{C} = .05 + .0333 + .025 = .1083$$

$$C = \frac{1}{.1083} = 9.23 \text{ MFD.}$$

"C" is used to represent farads of capacitance.

For any capacitor, the capacitance rating is directly proportional to the area of the capacitor plates and inversely proportional to the distance separating the plates. An additional factor in determining the value of capacitance is the material used for the dielectric. For instance, a given set of capacitor plates with a fixed separation will have a certain rating of farads with air as the dielectric but will have a higher value of capacitance if mica or ceramic or glass is substituted for air as the dielectric. Capacitance increases with increase in area of the plates and decreases as the plates are moved further apart.

If several capacitors are connected in series, the combined value of capacitance will be less than the capacitance value of the lowest rated capacitor in the group, in accordance with the formula,

$$\frac{1}{C} = \frac{1}{C_1} + \frac{1}{C_2} + \frac{1}{C_3} + \ldots$$

where C is the combined capacitance value in farads and C_1, C_2, C_3, etc. are the individual values of capacitance of each of the capacitors in the group. When a number of capacitors are connected in parallel, the combined capacitance value is equal to the numerical sum of the individual values of capacitance of the capacitors in the group, i.e. $C = C_1 + C_2 + C_3 + C_4 + \ldots$

Capacitive Reactance

As described previously, a capacitor connected in an alternating current circuit permits the current to flow by reason of its ability to store current on one direction of current flow, then discharge the current, then store current as the ac current flows in the opposite direction. But the capacitor does offer opposition to current flow. This opposition depends upon two factors: the value of capacitance and the rate at which the ap-

CAPACITORS IN PARALLEL

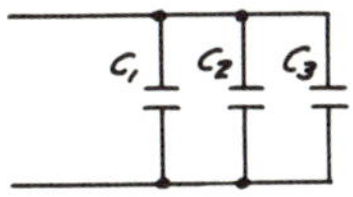

Total capacitance of circuit = C

$C = C_1 + C_2 + C_3$

EXAMPLE:

FIND C IF C_1 = 20 MFD, C_2 = 30 MFD, C_3 = 40 MFD.

SOLUTION:

$C = C_1 + C_2 + C_3$

$= 20 + 30 + 40 = \underline{\underline{90\ \text{MFD}}}$.

plied voltage is changing—which is the frequency of the alternating-current voltage source. With an ac voltage connected across only a capacitor, an increase in either the capacitance of the capacitor or an increase in the frequency of the ac voltage will decrease the opposition that the capacitor offers and will, as a result, cause an increase in the current flowing through the circuit. The opposition to current flow is called "capacitive reactance" and is inversely proportional to capacitance and frequency. Like resistance and inductive reactance, capacitive reactance is measured in ohms and has a value from the formula:

$$X_C = \frac{1}{2\pi fC}$$

where f = frequency of the voltage source in cycles per second,

π = 3.14 and

C = value of capacitance in farads.

When the capacitance is given in microfarads, the formula becomes:

$$X_C = \frac{1{,}000{,}000}{2\pi fC}$$

CAPACITIVE REACTANCE

The opposition offered by a capacitor to the flow of alternating current is called capacitive reactance and is measured in ohms.

$$X_C = \frac{1}{2\pi fC} \qquad I = \frac{E}{X_C}$$

X_C = capacitive reactance (ohms)
π = 3.14
f = frequency (cycles per sec.)
C = capacitance (farads)
I = effective current (amperes)
E = effective voltage (volts)

EXAMPLE: FIND I, I_2 AND I IN THE CIRCUIT BELOW IF C_1 = 200 MFD, C_2 = 100 MFD.

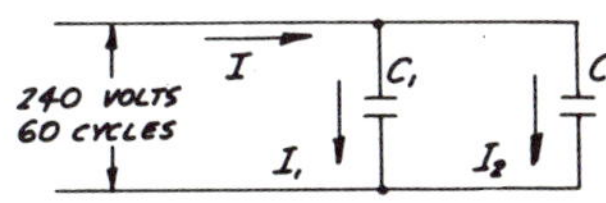

SOLUTION:

$X_{C_2} = \frac{1}{2\pi f C_2} = \frac{1\,000\,000}{(377)(100)} = 26.52$ OHMS.

$I_2 = \frac{E}{X_{C_2}} = \frac{240}{26.52} = \underline{\underline{9.05\ \text{AMPS}}}$.

$X_{C_1} = \frac{1}{2\pi f C_1} = \frac{1\,000\,000}{(377)(200)} = 13.26$ OHMS.

$I_1 = \frac{E}{X_{C_1}} = \frac{240}{13.26} = \underline{\underline{18.10\ \text{AMPS}}}$.

$I = I_1 + I_2 = \underline{\underline{27.15\ \text{AMPS}}}$

CHECK:

TOTAL CAPACITANCE $C = C_1 + C_2 = 300$ MFD.

$X_C = \frac{1}{2\pi fC} = \frac{1\,000\,000}{(377)(300)} = 8.84$ OHMS.

$I = \frac{240}{X_C} = \frac{240}{8.84} = \underline{\underline{27.15\ \text{AMPS}}}$.

As in the case of a circuit containing only inductance, a circuit containing only capacitance will have a current flow equal to

$$I = \frac{E}{X_C}$$

where I = effective or rms current,

E = rms voltage applied and

X_C = capacitive reactance in ohms.

Capacitance and Resistance

AS STATED previously, when an ac voltage is impressed across a pure capacitance, the current flow through the capacitance follows an ac wave pattern which leads the voltage wave by 90 electrical degrees. Or another way to say it is that the voltage lags the current by 90 degrees in a purely capacitive circuit. But because capacitors and other capacitive conditions are always concerned with circuits made up of wires and cables, the resistance of such wires and cables is added to the capacitive reactance so that there is no such thing as a purely capacitive circuit.

When a resistance and capacitance are connected in series across a voltage source, a single current flows with an in-phase relation to the voltage drop across the resistance and a 90° leading phase relation to the voltage drop across the capacitance.

These two voltage drops—the one across the resistance and the one across the capacitance—are therefore 90° out of phase with each other. As was also found in the case of resistance and inductance in series in a circuit, the simple mathematical sum of the two voltage drops is not equal to the source voltage impressed on the circuit.

The source voltage across the series load of resistance and capacitance is equal to the vector sum of the voltage drop across the resistance and the voltage drop across the capacitance in accordance with the following formula:

$E_S = \sqrt{E^2_R + E^2_C}$ where
E_S = voltage impressed on circuit,
E_R = voltage drop across resistance, and
E_C = voltage drop across capacitance.

An example of the voltage relationships in a circuit containing resistance and capacitance in series is given in an accompanying illustration. Other variations on the basic formula for voltage are as follows:

$$E_S = \sqrt{(IR)^2 + (IX_C)^2}$$
$$= \sqrt{I^2(R^2 + X^2_C)}$$
$$= I\sqrt{R^2 + X^2_C}$$

Then from those variations on the voltage formula, we derive the following on current and impedance:

$$I = E_S \div \sqrt{R^2 + X^2_C} = E_S \div Z$$

therefore,

$$Z = \sqrt{R^2 + X^2_C}$$

This last formula indicates that the impedance to current flow represented by resistance in series with capacitance can be reduced to an ohmic value if the resistance and capacitive reactance are known. And the capacitive reactance can be readily found from the value of capacitance involved. And if the voltage impressed on the circuit is known, the current flow through the circuit can be readily found from the formula $I = E_S \div Z$. Then the voltage drop across the resistance can be found from $E = I \times R$. And the drop across the capacitance can be found from $E = I \times X_C$.

Circuit Reactance

The current wave through a capacitance leads the voltage across the capacitance by 90°. The current wave through an inductance lags the voltage across the inductance by 90°. When they are connected in series the effect of capacitance is, therefore, 180

RC SERIES CIRCUIT - VECTOR RELATIONSHIPS

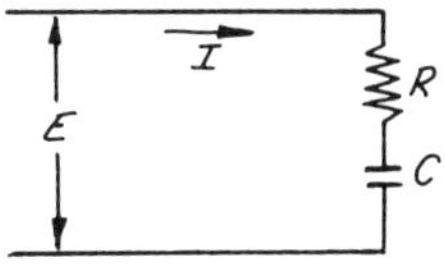

Voltage E_R across resistor R is in phase with current I. Voltage E_C across capacitor lags current by 90°. Supply voltage E lags current by an angle θ due to presence of capacitor in circuit.

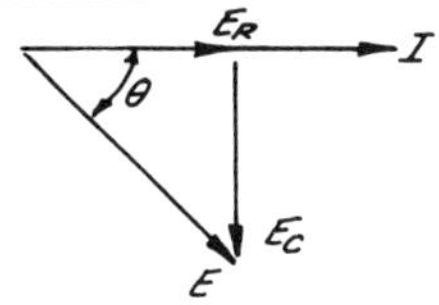

degrees out of phase with the effect of inductance. This means their effects are exactly opposite and tend to neutralize or cancel out each other.

If an inductive reactance and a capacitive reactance are connected in series, the combined effect is equal to the mathematical difference between their ohmic values, and the resultant effect has the character of the larger of the two types of reactance. That is, if the inductive reactance is larger than the capacitive reactance, the net reactance will be inductive and the current will lag the voltage. Or vice versa. The combined effect of the two types of reactance is simply referred to as *reactance*. The formula is:

$X = X_L - X_C$ where
X = reactance in ohms,
X_L = inductive reactance in ohms, and
X_C = capacitive reactance in ohms.

Impedance

Of course, as pointed out before, any circuit which consists of inductance and capacitance in series will also contain resistance in series with the reactances. The combined effect of resistance and reactance is *impedance* (the total opposition offered by a circuit to current flow). The value of impedance for any circuit can be found from the following formula.

$$Z = \sqrt{R^2 + (X_L - X_C)^2}$$

The above formula can be used to determine impedance for any circuit—a circuit with resistance and only inductive reactance, a circuit with resistance and only capacitive reactance or a circuit with resistance and both inductive and capacitive reactance.

When one or the other reactance is not included in the above formula, it can be simply represented by a zero. And the appearance of a minus sign before X_C in the event that there is no inductive reactance or the capacitive reactance is a larger ohmic value than the inductive reactance poses no

RC CIRCUIT IMPEDANCE

In a series circuit containing resistance and capacitance,

$$Z = \sqrt{R^2 + X_C^2}$$

where Z = circuit impedance (ohms)
R = resistance (ohms)
X_C = capacitive reactance (ohms)

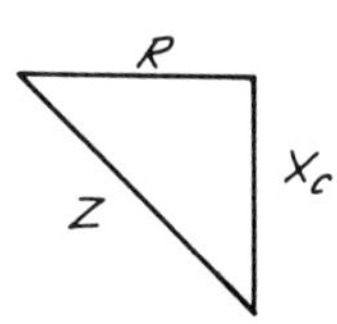

RLC SERIES CIRCUITS

Where both inductance and capacitance are included in the circuit, the net reactance X is given by

$$X = X_L - X_C$$

and the circuit impedance Z is given by

$$Z = \sqrt{R^2 + (X_L - X_C)^2} = \sqrt{R^2 + X^2}$$

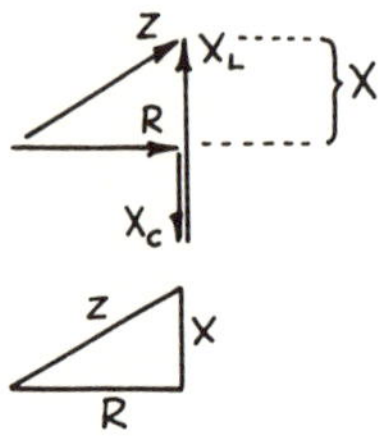

These diagrams illustrate this relation graphically. Resistance is drawn to scale horizontally; X_C and X_L are drawn at right angles to R and 180° out of phase with each other. The net reactance X is shown where X_L is greater than X_C and vice versa.

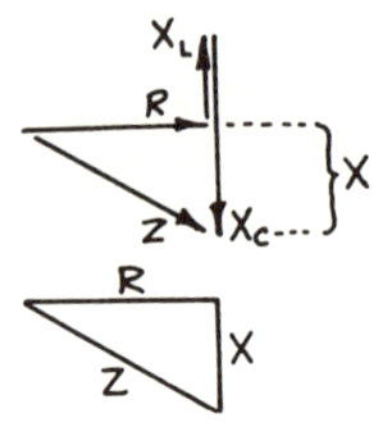

EXAMPLE:

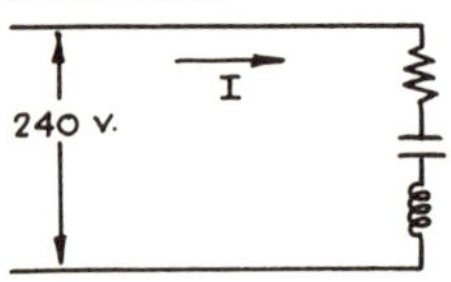

R = 50 OHMS
X_C = 20 OHMS
X_L = 40 OHMS

FIND IMPEDANCE Z, EFFECTIVE CURRENT I, AND VOLTAGE DROPS ACROSS R, X_C AND X_L.

SOLUTION:

$$Z = \sqrt{R^2 + (X_L - X_C)^2} = \sqrt{(50)^2 + (40-20)^2}$$

$$= \sqrt{2500 + 400} = \sqrt{2900} = 53.85 \text{ OHMS}$$

$$I = \frac{E}{Z} = \frac{240}{53.85} = 4.46 \text{ AMPS}$$

$$E_R = IR = (4.46)(50) = 223 \text{ VOLTS}$$

$$E_C = IX_C = (4.46)(20) = 89.2 \text{ VOLTS}$$

$$E_L = IX_L = (4.46)(40) = 178.4 \text{ VOLTS}$$

$$E = \sqrt{E_R^2 + (E_L - E_C)^2} = 240 \text{ VOLTS}$$

problem because a minus number squared is a positive number.

Practical application of the theory of ac circuit operation and analysis of circuit impedance characteristics is a vital part of modern electrical design and construction of systems for power, light and heat.

Given conditions usually include the circuit voltage, resistance and inductance of circuit conductors and the load characteristics of utilization devices—wattages of incandescent lamps, volt-amperes of fluorescent lamps, full-load currents drawn by motors and other electrical coil devices, and wattages of electric heating equipment.

Sizing and layout of circuit conductors, including voltage drop calculations and circuit power loss study, depend upon a sound understanding of ac theory on resistance, reactance and impedance and phase relationships.

Accompanying diagrams show the vector relationships involved with resistance, reactance, impedance, current and voltage in single-phase series circuits.

In electrical systems for power and light, capacitance is usually an accidental factor contributed by the insulation on wires in conduit. The insulation acts like the dielectric, the conductor is one plate of the capacitor and the (grounded) conduit is the other plate of the capacitor. At voltages under 600 volts, this capacitive effect is negligible.

Capacitance becomes a consideration when capacitors are added to a circuit to offset the effect of lagging current due to inductive elements.

Single-Phase Circuits

THE POWER in any ac circuit is a function of current, voltage and the phase angle, if any, between the current wave and the voltage wave. This differs from dc circuits, in which power is simply equal to the current value times the voltage value, with the product expressed as watts.

In an ac circuit which contains only a resistive load, the current and voltage waves are in phase and the power at any instant is equal to the instantaneous value of current multiplied by the instantaneous value of voltage at the same instant with the product expressed as watts. However, a quicker way to express the power is to average the power over one half cycle. Thus, the average power in each half cycle is equal to the rms value of current times the rms value of voltage. As a formula, this is basically the same as the power formula for dc—

$P = E \times I$, where

E=rms volts, and
I=rms amperes.

When an ac circuit contains reactance in addition to resistance, the current and voltage waves have a phase difference between them. It does not matter whether the reactance is inductive only or capacitive only (or is the difference between inductive and capacitive reactance in the same circuit). So long as the circuit has a net value of reactance, there will be some phase angle, with the current either leading or lagging the voltage, depending upon the nature of the reactance. In such circuits, the power in the circuit is *not* simply the product of voltage and current.

When the current in a circuit is out of phase with the voltage, there are regularly recurring periods when the voltage is positive and the current is negative. Under such conditions, the product of positive voltage and negative current in each cycle is negative, indicating that the circuit is delivering power to the source of energy supply during each cycle.

• **Reactance Only**—When the current and voltage waves of ac have a 90° phase difference between them, it can be shown that during one cycle the product of instantaneous values of current and voltage will produce equal amounts of both negative and positive power. That is, during part of the cycle, the energy source applied to the circuit delivers power into the circuit (the positive power); and during another part of each cycle, the circuit returns the same amount of power to the source.

Taking an inductance coil, for instance, during each cycle of the ac, power is twice delivered to the coil and is twice returned by the coil to the source. The net result of this action is that the power expended in the

POWER *and* POWER FACTOR

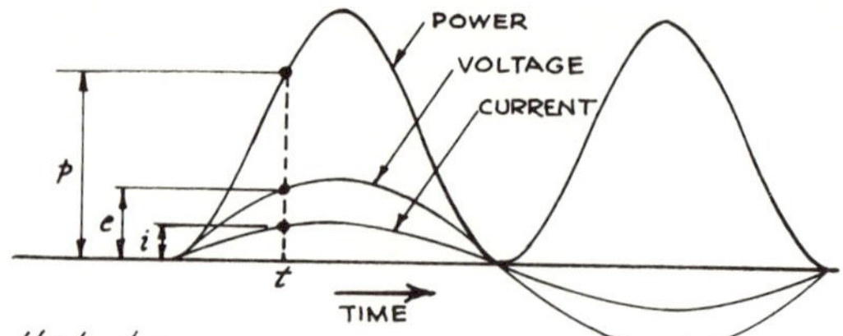

The product of the current i and the voltage e at any instant of time t gives the power p at that instant. Note that when both e and i are negative, their product, and thus p, is positive. Therefore power is being expended throughout the cycle.

In terms of effective values,

$P = EI$

when current and voltage are in phase.

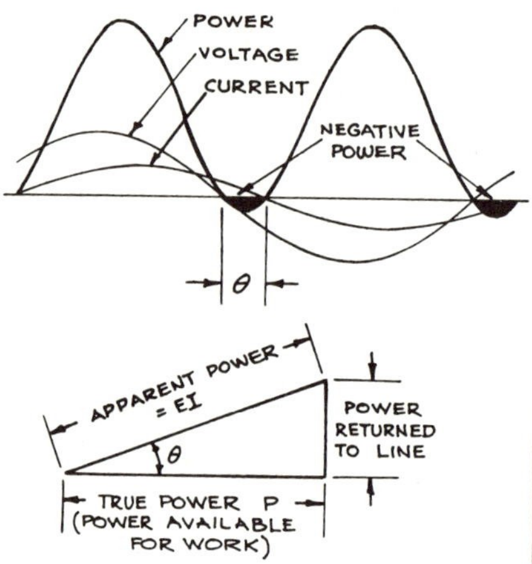

EXAMPLE: *A current of 7 amps lags a voltage of 240 volts by 30°. What is the power factor and the true power taken by the load?*

FROM TRIGONOMETRIC TABLES,

COS 30° = 0.866

THEREFORE, POWER FACTOR = 0.866 OR 86.6 %.

TRUE POWER = EI COS θ
= (240)(7)(0.866)
≅ 1455 WATTS

circuit is zero when the load is either a pure inductance or a pure capacitance. In such circuits, when voltage is applied, current flows and has a value determined from I=E/X, but the product of I and E is not watts. A purely reactive load merely exchanges power back and forth with the source, hence the use of the word "reactive" to describe such a load.

• **Reactance and Resistance** — All practical circuits contain resistance and may or may not contain reactance. Circuits in electrical systems for power and light almost never contain series capacitance but they do frequently contain series inductance—such as motor windings, ballasts, solenoids and other coil devices. When a circuit contains resistance and inductance, the current wave is out of phase with the voltage, lagging it by some phase angle between 0 and 90°. In such a circuit, the average power is not simply E x I but can be found from the formula P=E x I x cos Θ, in which Θ is the angle by which the current wave lags the voltage. The cosine of the angle Θ is called the "power factor" of the circuit and is a measure of the amount of power which is being consumed by the resistive part of the load, taking into account the effect of the inductance on the circuit.

In a series circuit containing resistance and inductance, the opposition to current flow is provided by the total impedance in accordance with $Z = \sqrt{R^2 + X^2_L}$. The current follows from I=E/Z. the apparent power delivered into the combined load is the circuit voltage times the circuit current, E x I. But the only part of the load consuming power is the resistance. **The inductance consumes no** power. The power consumed in the resistance is equal to the voltage across the resistance times the current through the resistance; or, since the voltage across the resistive part of the load is equal to I x R, the power in the resistance is equal to **I x I x R**—which is I^2 R. This is the true power being consumed in the circuit.

The cosine of the phase angle of a circuit is the power factor of the circuit and is the factor which determines what portion of the "apparent power" is "true power". The cosine of the angle will vary from 1, when the phase angle is zero and the load is purely resistive, to 0, when the phase angle is 90° due to a purely reactive load. The words "Power Factor" or simply the letters "PF" are commonly used instead of "cos Θ" to designate power factor. As a result, the formula for power in reactive circuits is often given as P = E x I x Power Factor. By transposing this formula, it can be seen that Power Factor =P ÷ (E x I) or true power ÷ apparent power.

When Θ is zero, the formula becomes P = E x I x 1, or P = E x I, the formula for in-phase operation. When Θ is 90°, P = E x I x 0 = 0, indicating that no power is being considered in a purely ~~relative~~ reactive load.

Power factor may be expressed as a decimal, having the exact numerical value of the cosine of the phase angle, e.g., a power factor of 0.7, a power factor of 1.0 (the maximum), etc. Or the decimal value of the cosine may be converted to percentage, such as a power factor of 70% or of 95% or of 100%.

• **Unity power factor** is another expression for the condition where the power factor is 1.0 or 100%, that is, the current and voltage are in phase.

In circuits with a phase angle between current and voltage, the simple product of voltage times current (which was described above as "apparent watts") is most commonly called the "volt amperes" rating of the circuit. The kva (kilovolt-ampere) rating of a circuit or device is equal to voltage times current divided by 1000. A value of volt-amperes or kilovolt-amperes for a circuit can be multiplied by the power factor of the circuit to determine the watts or kilowatts of power being consumed in the circuit.

R-L CIRCUIT

In a series circuit containing resistance and inductance, the current lags the voltage by an angle θ. During part of each cycle, the voltage is negative while the current is positive; therefore their product is negative. This "negative" power is not available for work; it is actually power returned to the line. This may be represented by a right triangle, as shown by the drawing at the left. From trigonometry,

TRUE POWER P = EI COS θ,

where all values are effective values. The factor Cos θ is also known as the "power factor."

EXAMPLE: *A motor rated 240 volts, 9 amps draws 1728 watts at full load. What is its power factor?*

$$P = EI \cos\theta$$

$$1728 = (240)(9)(\cos\theta)$$

$$\cos\theta = \frac{1728}{(240)(9)} = \frac{1728}{2160}$$

POWER FACTOR = 0.8 OR 80%

Parallel Circuits

THUS FAR, discussion of ac circuits has covered single-phase series circuits. Such circuits are certainly common in electrical systems for power, light, heat, signals and communications. Two wires feeding an incandescent lamp constitute a series circuit with a resistive load. Two wires supplying a motor or solenoid coil make up a series circuit with a load impedance combining both resistance and inductance.

Operation and calculations of such circuits have been covered. But practical electrical systems are always made up of both series circuits and *parallel circuits.*

The basic approach to analysis of parallel circuits is to recognize that the voltage impressed across each of the loads in parallel is the same and that the current through each load will be determined by the size and nature of each load—how much resistance and/or reactance is each. Then the currents through the individual branches must be combined to obtain the resultant current of the parallel combination.

It should be noted that there is a set of opposite relationships between a series circuit and a parallel circuit. In a series circuit, there is one value of current through the series loads, the voltage across any one load is equal to the current multiplied by the impedance of the load and the voltage impressed on the circuit is equal to the vector sum of the voltages across the series loads. In a parallel circuit, there is one value of voltage across all of the loads connected in parallel, the current through any one load is equal to the voltage divided by the impedance of the load and the total current of the parallel circuit is equal to the vector sum of the currents through the parallel loads.

In practical electrical systems, branch circuits from lighting and appliance panelboards are parallel circuits. The voltage across the two wires of the circuit (120 volts in the typical case) is applied across a number of lighting fixtures and/or appliances connected in parallel. The current which the circuit must supply (commonly called the "load" on the circuit) is the vector sum of the currents drawn by the individual lighting units and/or appliances.

In the same way, the feeder to this lighting and appliance panelboard makes up a parallel circuit which has a given voltage across the branch circuits and the total current of the feeder parallel circuit is the vector sum of the currents drawn by the individual branch circuits. And similarly, the feeder to a switchboard consists of circuit with the sub-feeders in parallel. Thus it can be readily seen that any system is a tree-like combination of series and parallel circuits.

As described previously under dc circuits, if a number of resistances are connected in parallel across a source of voltage, each resistance will draw current in accordance with $I = E/R$ and the total current drawn by the parallel combination will be simply the arithmetical sum of the individual load currents (which is the vector sum when all of the currents have the same phase relationship).

For example, if five lamps, each having a resistance of 120 ohms, are

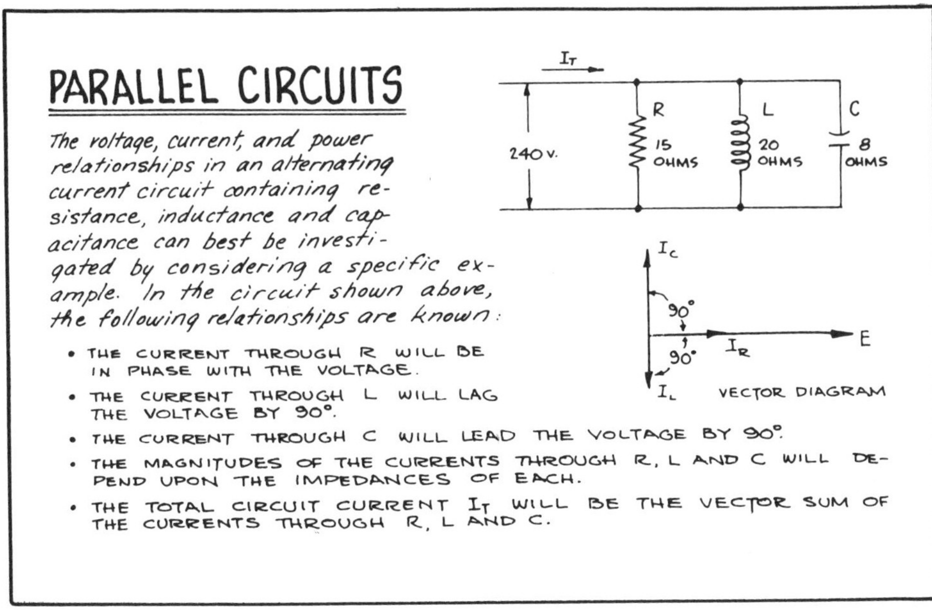

connected to a 120-volt branch circuit (using standard parallel connection), each lamp will draw 1-amp current ($I = E/R = 120/120$) and the total current load on the circuit will be 5 amps. But when the parallel loads combine resistance and reactance, phase angles must be considered; and solution of the circuit is more complicated.

Consider first a voltage impressed across a resistance connected in parallel with an inductance. The current through the resistance will be equal to E/R and will be in phase with the voltage. The current through the inductance will be equal to E/X_L and will be lagging the voltage by 90°. So far, we have treated each load as a series circuit. Now the two values of current obtained can be vectorially combined to find the resultant current. This is readily found from the following formula (which is basically the same as the formula previously used to obtain the vector sum of two voltages with a 90° phase difference between them):

$I_T = \sqrt{I_R^2 + I_L^2}$, where

I_T = total current of parallel circuit,
I_R = current drawn by resistance, and
I_L = current drawn by inductance.

It should be noted that the two currents of the parallel loads are 90° out of phase with each other.

If a resistance and capacitance are connected in parallel, a similar analysis is made. The currents through the loads will again be 90° out of phase, but this time the current through the capacitance will be leading the current through the resistance. But the resultant current is found from the same formula:

$I_T = \sqrt{I_R^2 + I_C^2}$, where
I_C = current drawn by capacitance.

Another example of the parallel circuit consists of resistance, inductance and capacitance in parallel. In such a circuit, the current is first found for each of the parallel loads. Then, because the current through the inductance is lagging the impressed voltage by 90° and the current through the capacitance is leading the impressed voltage by 90°, these two currents are 180° out of phase or, as it is some-

CURRENT FLOW:

CURRENT THROUGH R

$$I_R = \frac{E}{R} = \frac{240}{15} = \underline{16 \text{ AMPS.}}$$

CURRENT THROUGH L

$$I_L = \frac{E}{X_L} = \frac{240}{20} = \underline{12 \text{ AMPS.}}$$

CURRENT THROUGH C

$$I_C = \frac{E}{X_C} = \frac{240}{8} = \underline{30 \text{ AMPS.}}$$

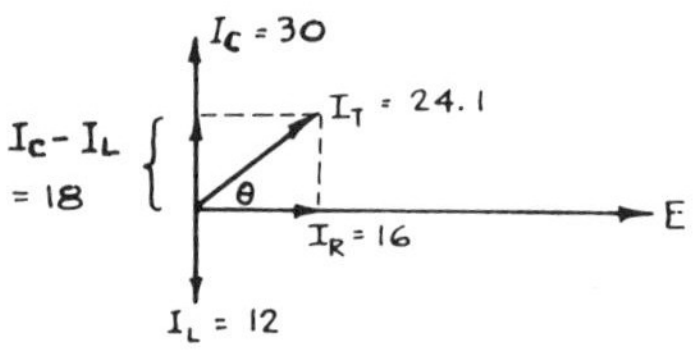

CIRCUIT CURRENT I_T

I_C AND I_L ARE 180° APART, OR DIRECTLY OPPOSITE; THEREFORE THEY CANCEL EACH OTHER TO A CERTAIN EXTENT:

$$I_C - I_L = 30 - 12 = 18$$

THERE IS THUS A RESULTANT 18 AMPS OF CAPACITIVE CURRENT, AS SHOWN IN THE VECTOR DIAGRAM AT THE LEFT.

THE RESULTANT OF THIS CURRENT AND I_R IS FOUND AS FOLLOWS:

$$I_T = \sqrt{(18)^2 + (16)^2}$$
$$= \sqrt{324 + 256}$$
$$= \sqrt{580}$$
$$= \underline{24.1 \text{ AMPS}}$$

POWER FACTOR:

$$\cos\theta = \frac{\text{ADJACENT SIDE}}{\text{HYPOTENUSE}}$$
$$= \frac{16}{24.1} = 0.664$$

POWER FACTOR = $\underline{66.4\ \%}$

POWER:

TRUE POWER TAKEN BY COMBINED LOAD OF R, L AND C:

$$P = EI\cos\theta$$
$$= (240)(24.1)(.664)$$
$$= \underline{3841 \text{ WATTS}}$$

times stated, are in phase-opposition.

The resultant of two such currents is the arithmetical difference between them. Then this difference current is combined with the current through the resistance load, in accordance with either of the above formulas, to find the resultant current of the parallel circuit.

In parallel circuits which include reactance of either type, the resultant current drawn by the parallel load will not be in-phase with the voltage impressed on the circuit. The resultant current will lead or lag the circuit voltage depending upon the type and amount of net reactance of the circuit—inductive or capacitive—and the value of the resistance. As a result, the power consumed in such a circuit will *not* be equal to voltage times current. Instead, just as in the case of phase difference between current and voltage in a series circuit, the power can be found from the formula—

$P = E \text{ x } I \text{ x } \cos\theta$, where
P = power in watts,
E = circuit voltage,
I = resultant current, and
θ = phase angle of circuit.

The cos θ—which is the power factor of the circuit—is equal to $I_R \div I_T$, the current through the resistive load divided by the resultant current. The simple product of E x I is the volt-ampere value of the circuit, the apparent power of the parallel load.

Complex Circuits

SO FAR, single-phase circuits have been considered as series or parallel connections of resistances, inductances or capacitances.

In the case of parallel circuits, idealistic conditions were assumed in which a pure resistance was connected in parallel with a pure inductance and/or in parallel with a pure capacitance. Although such conditions are possible in practical circuits, it is much more common to find, say, a pure resistance in parallel with another load made up of a resistance and inductance in series. Or, a capacitance might be connected in parallel with a load made up of resistance and inductance in series. Such circuits are actually combinations of series and parallel circuits.

A typical example of a complex circuit in common use would be a 120-volt branch circuit supplying an incandescent lamp load and a motor load. The loads would be connected in parallel. The lamp load would be a pure resistance and the motor load would be represented by a series connection of the resistance and inductance of the motor coil windings. Such a circuit could be one lamp in parallel with one motor, or it might be several lamps and a number of motors—all connected in parallel.

Another example of a complex circuit would be a panelboard, in which branch circuits are connected in parallel to the busbars in the panel. Each branch circuit could be reduced to a single equivalent load, and then all of these loads are considered connected in parallel. For instance, a branch circuit with 10 lamps connected from phase to neutral and each drawing 1 amp of current can be readily converted to a single-resistance load drawing 10 x 1 or 10 amps. And at 120 volts across this load, the effective resistance of the total load is 120/10 or 12 ohms. Then the effective resistances of all of the branch circuits supplying only resistive lamp loads can be combined, as parallel resistances, into a single resistive load connected from phase to neutral across the feeder to the panel.

Then, if the panel has branch circuits which supply motor loads or motor-plus-lighting loads, these loads can be successively treated as series and parallel circuits. From this, a single equivalent series load of resistance plus inductance can be obtained. Then this series load is combined in parallel with the purely resistive lamp load determined previously. From this final parallel hookup, we can determine the current, total load impedance characteristics and power factor of the 120-volt feeder to the panel.

Basic analysis of a complex circuit can be made by considering a pure resistance connected in parallel with a series load of resistance and inductance. With a known voltage across the two parallel loads, the current through the pure resistance can be determined from $I = E/R$, and this current is in phase with the voltage. Then the series load of resistance and inductance must be considered by it-

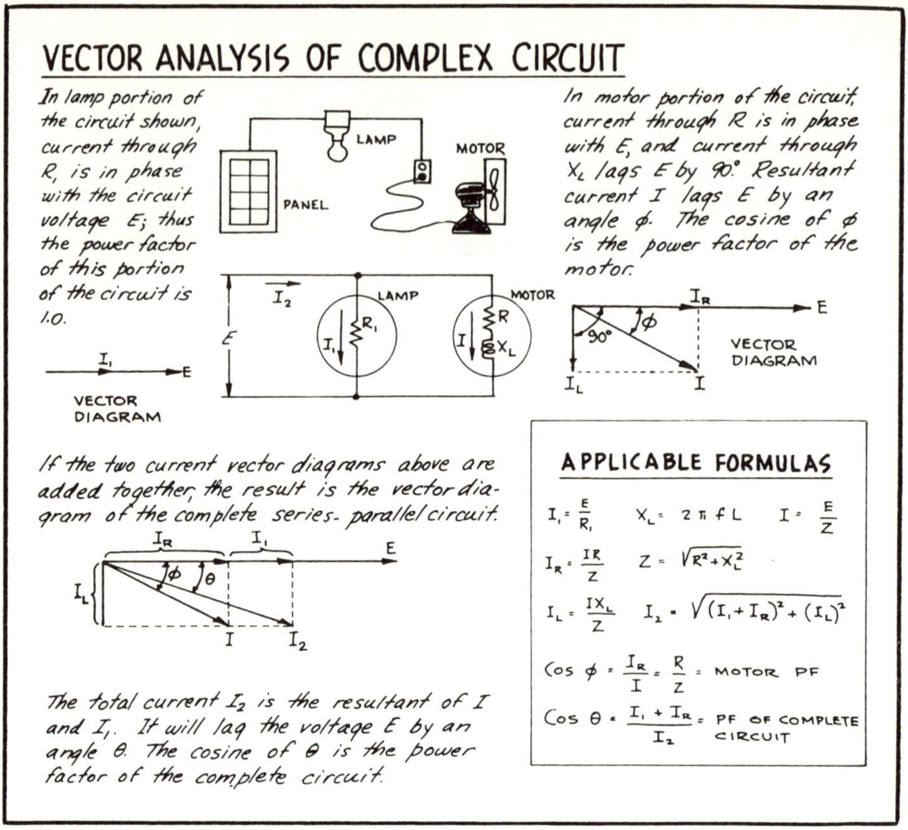

self. The inductance is converted to inductive reactance from $X_L = 2\pi fL$.

The impedance of the series load is obtained from $Z = \sqrt{R^2 + X_L{}^2}$. Then the current through the series load is E/Z. And the power factor of the series load circuit is R/Z, representing the value of the cosine of the angle by which the current through the series load lags the voltage impressed across the load.

The above parallel circuit is solved by vector analysis, observing basic geometric relationships. The voltage vector, which is common to both of the parallel current paths, is laid off to the right, with the current vector for the purely resistive load in phase with the voltage vector and the current vector for the series R-plus-L load lagging the voltage vector by the appropriate phase angle, as indicted by the power factor of the R-plus-L load. The two current vectors are then vectorially combined to find the resultant current drawn by the parallel combination. And the cosine of the angle by which this resultant current lags the impressed voltage is the power factor of the complete parallel circuit. The effective impedance of the parallel circuit is obtained by dividing the impressed voltage by the resultant current.

A common application of the parallel circuit is that of a power factor correcting capacitor connected in parallel with a series R-plus-L load.

When a load connected across circuit conductors consists of resistance and inductance in series—such as a single motor on a branch circuit, or the effective load of motors and lighting on a branch or feeder circuit—the current will lag the voltage by some value, depending upon the ratio of R to L. That is, the power factor will be less than unity and the product of voltage times current (voltamperes) will be greater than I^2R (the wattage of the load). Part of the current being drawn is magnetizing current and not useful current. The lower the power factor of the load, the greater will be the magnetizing current.

Thus a number of low-power-factor induction motors on a system can represent a sizable magnetizing current.

Because a capacitor draws leading current, the right size of capacitor can be used in parallel with the series R-plus-L to provide the magnetizing current to the load. In doing this a capacitor cancels the reactive component of the current drawn by the series load. The reactive component of the series load current and the purely reactive current drawn by the capacitor are 180° out of phase (90° leading and 90° lagging). By eliminating the reactive part of the load current, the capacitor reduces the total current drawn from the line to supply the given amount of watts load.

When a capacitor corrects the circuit power factor to unity, the product of volts times amps *does* equal the wattage of the load. And the reduction in line current produced by power factor correction reduces line voltage drop and, consequently, power loss and increases the ability of conductors to supply real power to loads.

It is important to understand that when the line current drawn by a series R-plus-L load is reduced by connection of a PF correcting capacitor, the current through the series load remains unchanged. It still takes the same current through it as it took before the capacitor was applied. But the reactive part of the current to the series load is being supplied by the capacitor and the line only has to supply the active component of current for the resistive part of the series load.

POWER FACTOR CORRECTION

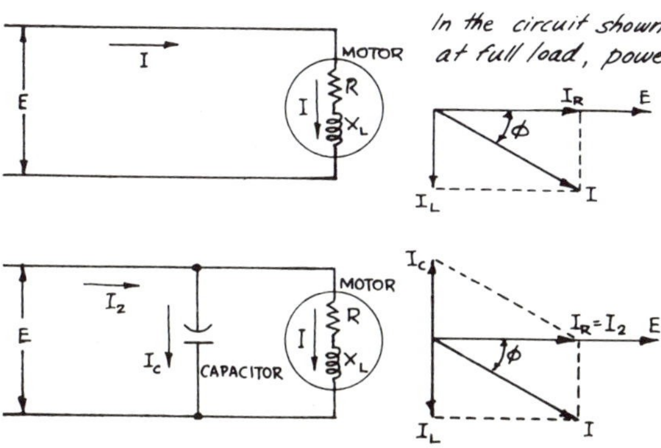

In the circuit shown at left, assume $I = 143$ amps at full load, power factor = 0.7. Current I lags the voltage E by an angle ϕ; $\cos\phi = 0.7$. $I_R = I\cos\phi = 143 \times 0.7 = 100$ amps.

If a capacitor is connected across the motor as shown to raise the circuit power factor to 1.0, the overall power factor angle $\theta = 0$ and the current I_2 furnished by the line is equal to I_R, or 100 amps.

Three-Phase Circuits

ALTHOUGH single-phase ac circuits are widely used and are involved in just about every electrical system, generation and distribution of ac electric power is most commonly done with three-phase circuits.

Large utility generators are generally three-phase machines, transmission is made over three-phase lines and most industrial and commercial electrical systems in buildings use three-phase circuits for power distribution.

Three-phase circuits require less weight of conductors than single-phase circuits of the same power rating. And three-phase motors are generally smaller, lighter in weight and more efficient than single-phase motors of the same horsepower. Overall, three-phase circuits are used because they are more efficient than single-phase.

As shown in an accompanying diagram; three-phase power is basically derived from a generator with three separate windings. These windings are spaced around the circular cross section of the rotating cylinder, with each winding taking up one third of the circumference of the rotor. Thus, each winding takes up ⅓ of the 360 degrees of the circumference. This means each winding takes up 120 degrees, and any turn of one winding is physically displaced 120 degrees from the same turn on either of the other two windings.

When the rotor is rotated, the coil windings on the rotor are made to cut through the magnetic lines of flux produced by the field poles of the generator. A sinusoidal voltage is produced in each of the three windings. But because the coils are displaced 120 degrees from each other, the voltage wave generated in any one coil will be 120 degrees ahead of the wave in one of the other coils and 120 degrees behind that in the third coil.

Assuming counterclockwise rotation of the rotor, it can readily be seen that each coil passes through the flux lines at a different time. As explained previously, such difference in time is referred to as a phase difference. The three voltage waves produced are, therefore, 120 degrees out of phase with each other. Designating one coil "A" and the one that immediately follows it in rotation as "B" and the one that follows B as "C", the phase sequence of voltages is A, B, C.

There are two basic connection arrangements used for three-phase systems. One is called the "wye" connection and the other is called the "delta" connection. Voltage sources (generators and transformer secondaries) and load devices may be connected either in wye or delta.

Wye Connection

As shown, the ends of the three windings are designated, a, a', b, b', c

3-PHASE GENERATOR. *Wye-(or Star-) connected rotor is shown feeding a wye-connected load. Simplified diagram indicates internal connections to neutral point and slip rings.*

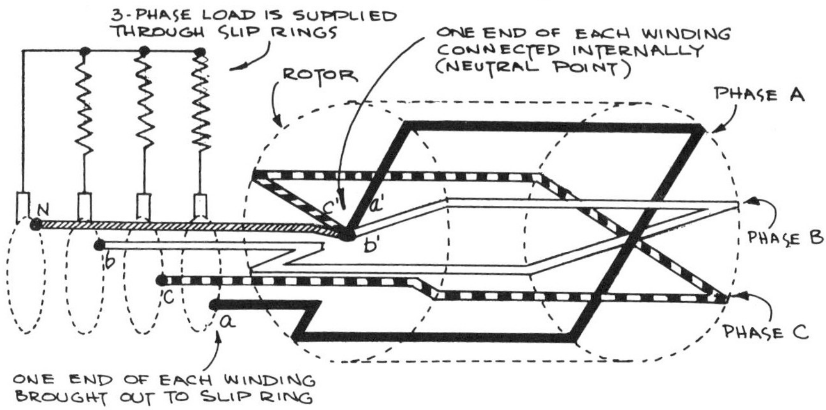

and c'. If the ends marked a', b' and c' are connected together, the voltage relationships remain the same in the windings. The connection produces a source of ac voltage which is called "three-phase, wye connected." The wires of this system are brought out of the generator—one wire is brought from each of the individual ends of the windings (ends a, b and c) and one wire may or may not be brought out from the common point connection of a', b' and c'. Although the generator described here has the voltage produced in the windings on the rotor, with slip rings required to bring the leads out, the magnetic field could be placed on the rotor and the generated voltage produced in windings around the stationary frame of the generator.

The common point of the generator windings (the connection of a', b' and c') is called the neutral point of the wye connection. The word "wye" is taken from the layout of the voltage vectors which represent this type of connection for study purposes. The three vectors are laid out on paper with 120° between each pair and with all vectors originating from the common or neutral point. The vector layout forms the letter "Y". Each conductor from one of the outer ends of the Y (usually referred to as "wye") is called a phase conductor. The conductor from the neutral point is called the neutral conductor.

Typical generators used by the electric utility companies generate three-phase voltage at high levels, such as 13,000 volts. Through the use of transformers, these voltage levels are generally stepped up for transmission and distribution and stepped down for utilization purposes. But throughout transmission, distribution and utilization, certain relationships among the voltage phases are retained. For instance, when the voltage windings are connected in a wye, two levels of voltage are available from the connection. One voltage level is the voltage generated in each phase winding, the voltage from any one of the phase conductors to the neutral point. The other voltage level is the voltage between any pair of phase conductors.

If, for instance, a wye-connected generator produced 8,000 volts in each set of phase windings, that much volt-

REVOLVING FIELD generator requires only two slip rings to carry direct current to field poles. Stationary armature (stator) windings are connected directly to load circuit conductors. Sketch shows 8-pole rotor.

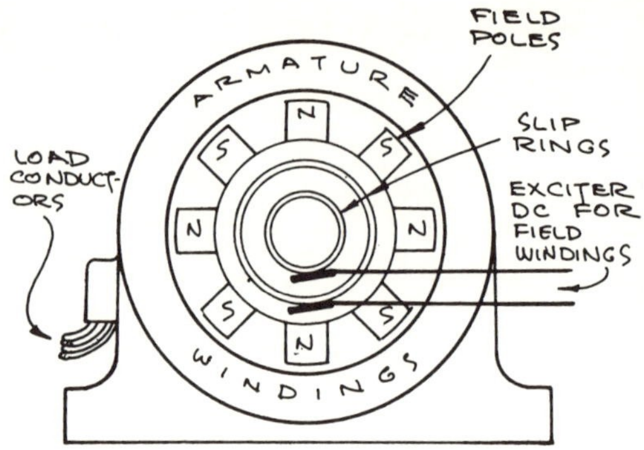

age would be measured between any phase conductor and the neutral of the generator. But the voltage between any pair of phase conductors would be equal to 1.73 x 8,000 or 13,840 volts. This latter voltage is the result of combining two phase-to-neutral voltages which are themselves 120° out of phase. Such a generator would be described as a 13,840Y/8,000 volt, three-phase, four-wire generator.

It should be noted that each voltage level in a wye system has one value, but that one value applies to each of three different phases. That is, taking the generator discussed above, there are three phases of 13,840 volts and three phases of 8,000 volts.

The voltage characteristics described here for a four-wire wye generator also apply to transformer connections made in wye, and to wiring systems themselves when supplied by wye connected voltage sources—generator or transformer. There is a given value of voltage between any pair of phase conductors, and another value of voltage between each phase wire and the neutral. Phase-to-phase voltage is equal to 1.73 x phase-to-neutral voltage, and phase-to-neutral voltage is equal to phase-to-phase voltage ÷ 1.73.

Common secondary distribution and utilization systems based on the three-phase, four-wire, wye arrangement are the 120/208-volt system and the 480/277-volt system. In the former, 120 volts can be obtained between the neutral conductor and any one of the three phase wires, and 208 volts exists between any pair of phase wires. In the latter system, 480 is the phase-to-phase voltage and 277 is the phase-to-neutral voltage. Such systems are generally obtained from transformers, or sometimes produced by generators.

WHEN a voltage source (generator or transformer secondary) is connected in wye, a conductor may or may not be brought out from the neutral point of the wye.

If single-phase load devices with voltage rating equal to the phase-to-neutral voltage of the wye are to be supplied, the fourth wire—the neutral wire—must be made available to obtain the required voltage. If, for instance, a 480Y/277-volt transformer secondary is to be used to supply 277-volt fluorescent ballasts, the neutral must be brought out of the transformer. But, if the same transformer secondary had to supply only three-phase motors and, say, one or more three-phase transformers for stepping down from 480 volts to 120/208 volts, three-phase, four-wire, there would be no need for bringing the neutral out of the transformer.

In this latter case, only the three phase wires from the outer ends of the wye are needed to supply 480 volts phase-to-phase to the motors and to the primaries of the step-down transformers. And that is true whether wye or delta connection is used in the motors. The neutral in a wye system is brought out only when needed.

Currents in Wye

In a wye-connected generator, when the output leads are connected to a load, current will flow through each phase winding into the line conductor connected to the winding, then through the external circuit and the load and back to the generator winding. This current flow will take place either with or without a neutral conductor, depending upon the type of load on the generator.

The important relationship to remember is that the current in any phase conductor from the generator to the load is equal to the current flowing through the corresponding phase winding on the generator. This is stated: I_L (line current)$=I_P$ (phase winding current). The same relationship holds for transformer windings.

In the common applications of today's electrical systems, wye-connected sources are generally transformer secondaries. The common loads fed from such sources are single-phase devices rated for the phase-to-neutral voltage (incandescent lamps, appliances, fluorescent lamps, single-phase motors) and three-phase devices rated for the phase-to-phase voltage (three-phase motors and three-phase step-down transformers). When three-phase motors are fed from the wye source, the normal current flow is the same in all three line conductors and the neutral is not used. That is called a balanced load. But when a wye source feeds a number of single-phase loads connected from phase to neutral, the loads may or may not represent balanced loading.

For instance, on a 120/208-volt, 4-wire wye system, two lighting fixtures might be connected from phase A to neutral, five fixtures from phase B to neutral and four fixtures from phase C to neutral. If all of the lighting fixtures are the same rating and each draws 2 amps, the current for phase A line will be 4 amps; for phase B, 10 amps; and for phase C, 8 amps. The neutral conductor of the system will carry the vector sum of these three currents which are 120° apart. If, however, the loads were balanced on the three phases, so that each line conductor was carrying, say,

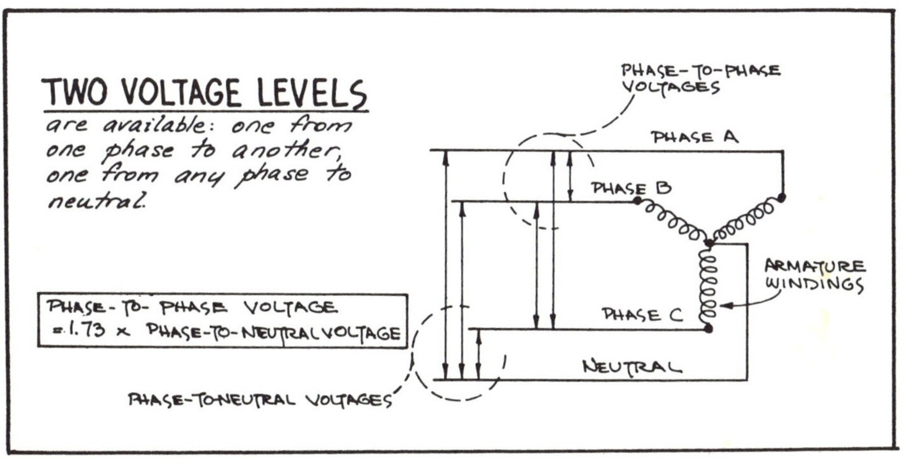

10 amps, the vector sum of these equal currents would be zero and the neutral would carry no current.

Of course, even with balanced loading on the three phases of a four-wire wye system, there is the possibility that not all of the loads will be operating at the same time. Thus there will be an operating unbalance and the neutral will carry current.

In determining currents in the line wires of a four-wire wye system, each phase can be treated as a separate single-phase circuit, but the neutral current must be determined as the vector sum of the three phase currents flowing into the neutral junction point of the load devices.

Note: With balanced loading on a 60-cycle ac four-wire wye system, the neutral wire carries no current. But it has been established by test that a balanced load of ballasts for fluorescent or mercury vapor lighting will produce third harmonic neutral current flow approximating phase currents in magnitude. There will be no 60-cycle current in the neutral but there will be 180-cycle current.

Delta Connection

Delta connection of the three 120°-displaced phase windings of the previously discussed generator can be made by connecting the windings end-to-end. In the basic generator described, the winding ends are connected as follows: a to c', c to b' and b to a'. Then a conductor is brought out from each of the junctions. And there is only one voltage value associated with each phase of the generator—that is the phase-to-phase voltage, or the "line voltage" as it is frequently called.

In the delta connection, the voltage of each of the generator phase windings is the same as the voltage between each pair of line conductors. As discussed previously, the voltage of the phase winding of the wye generator is the phase-to-neutral voltage, and the phase-to-phase or line voltage is 1.73 x the phase winding voltage. Delta connection, like wye connection, can be used for transformer as well as generator windings.

The delta connection gets its name from the triangular layout used for the three coils. If the vectors representing the magnitude and direction of voltage in the coils are connected end-to-end, they form a triangle which is the symbol for the Greek letter "delta." In diagrammatic form, delta connected generator and transformer coils are shown in a triangle layout.

Currents related to delta connected voltage sources involve two different values of current, similar to the way in which wye connections involve two different voltage levels. The current flowing in each phase winding (of

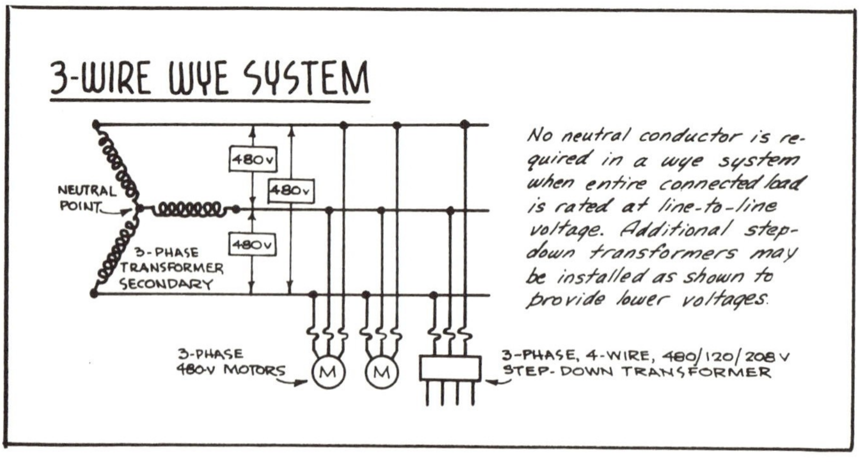

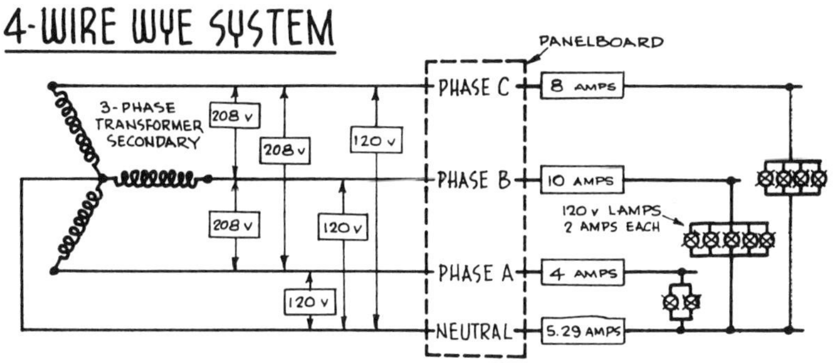

SUMMARY OF VOLTAGE-CURRENT RELATIONSHIPS, WYE-DELTA CONNECTIONS (BALANCED CONDITIONS)		
WYE	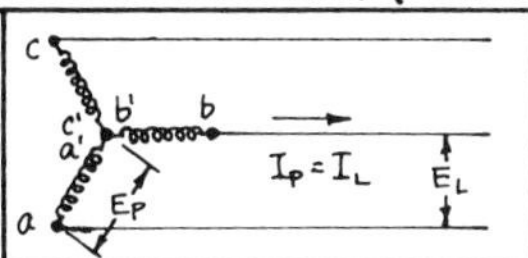	$I_L = I_P$ $E_L = 1.73\ E_P$
DELTA	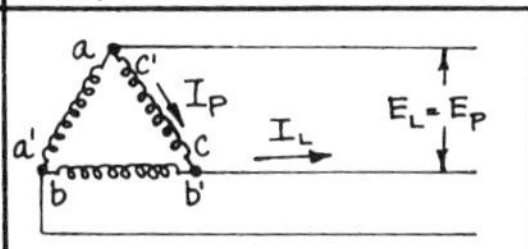	$I_L = 1.73\ I_P$ $E_L = E_P$
I_L = LINE CURRENT I_P = PHASE CURRENT	E_L = LINE VOLTAGE E_P = PHASE VOLTAGE	

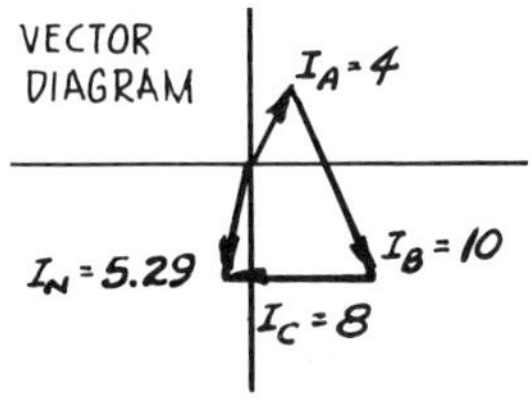

Unbalanced loading on 4-wire, 3-phase wye circuit results in a current flow in the neutral conductor. Magnitude of this current is the vector sum of the three phase currents.

the generator or transformer) has one value and is called the phase current. But since each line conductor from the source is connected at a junction of two of the phase windings, the current flowing in each line conductor from the source to the load is the vector sum of two phase currents. With balanced loading on the delta, the phase current will be the same in all three windings and the line current will be the same in all three line conductors. For any balanced delta connection, the relationship between phase and line current is—

$I_L = I_P \times 1.73$, where

I_L = line current, and
I_P = phase winding current.

BASED on foregoing discussion of wye- and delta-connected voltage sources, there are three basic considerations in analysis of three-phase electrical circuits:

1. The type of connection of the voltage source,
2. The makeup of the circuit conductors which connect the source to the load, and
3. The type of connection of the load to which the circuit delivers power.

The source—which, in building electrical systems, is most commonly a transformer secondary, but might occasionally be a generator—may be wye connected or delta connected. A wye source is generally used when single-phase load devices are to be supplied—such as incandescent and

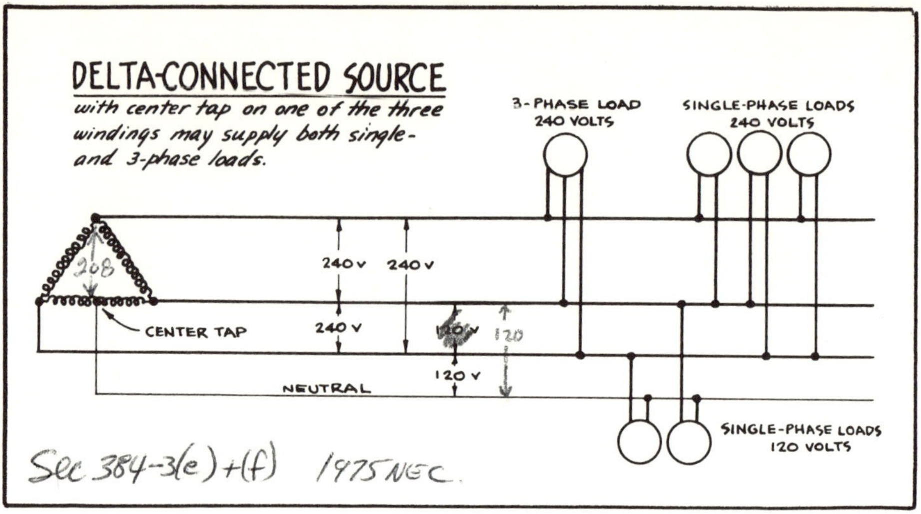

fluorescent lighting, appliances and small motors. The common systems used are 3-phase, 4-wire wye, 120/208 volts or 277/480 volts. In either system, the phase-to-neutral voltage is the one that is usually required for the single-phase loads, although some single-phase loads might be rated for the phase-to-phase voltage. But such a source also makes available the three-phase connections for three-phase motors and other three-phase load devices.

A wye-connected source might also be selected to serve purely three-phase loads, without using a neutral conductor, when the advantages of a grounded-neutral system are desired. When the neutral point of a wye transformer is connected to ground, any accidental ground on one of the phase conductors will cause fault current to flow and operate the circuit protective device (fuse or CB), opening the faulted circuit before a phase-to-phase fault might develop. A wye-connected transformer might also be used to supply only three-phase loads when the primary-to-secondary turns ratio of the transformer is such as to give the needed phase-to-phase voltage for the secondary load. Still another application for a wye-connected transformer without use of the neutral conductor is to supply single-phase loads which are rated for the phase-to-phase voltage of a wye secondary—such as 208-volt load devices on a 120/208-volt wye system or 480-volt single-phase loads on a 480/277-volt wye system.

Delta-connected sources are generally reserved for purely three-phase loads which require only three conductors to them. When a delta source is required to supply both single-phase lighting loads as well as three-phase loads, a four-wire system may be established by deriving a fourth conductor from the mid-point of one of the three transformer windings which make up the delta. Such a system is commonly called a "red leg" or "high leg" delta system. The usual voltage setup in such a system is 240 volts from phase to phase and 120 volts from the mid-phase (or neutral) conductor to each of the phase conductors connected to the winding ends. With this system, single-phase lighting and appliance loads are fed from the 120-volt terminals, and three-phase motor loads are fed from the phase conductors, although single-phase 240-volt loads could be fed by phase wires.

As noted previously, the number of conductors used to connect a voltage source to its load devices depends upon the load devices and how they are connected. To supply only three-phase motor loads, only three conductors are needed. A wye source can supply three wires to feed either a wye- or delta-connected motor. Or a delta source can also supply three wires to feed either a wye- or delta-connected motor. And the current and voltage relationships for wye and delta are the same for load connections as for source connections.

Four conductors are required for the supply circuit when a wye system is used with a neutral conductor or when a "red leg" delta system supplies a combination lighting and motor load. It should be noted that when a 4-wire wye distribution system is used to supply lighting and power loads, there is no need to use the neutral wire for connecting to the three-phase load devices. If, for instance, a 480/277-volt system is to supply a three-phase step-down transformer to provide 120/208 volts for lighting and receptacle outlets, only three wires have to be connected to a delta hookup of the transformer primary windings. Thus the 480 volts phase-to-phase of the wye are connected phase-to-phase

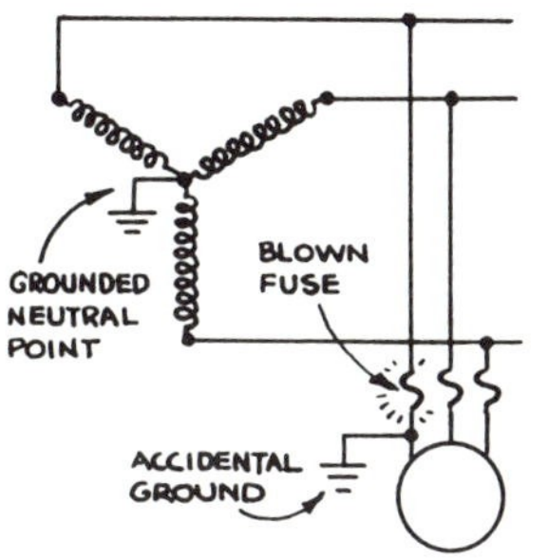

GROUNDED NEUTRAL POINT *of Y-connected source permits phase-to-ground fault detection, such as contact between phase conductor and raceway, thus eliminating shock hazard of energized raceway and other equipment.*

POWER IN 3-PHASE CIRCUITS

For both Y- and delta-connected systems:

$$P = \sqrt{3}\, E_L I_L\, PF$$

where P = TOTAL POWER (WATTS)
$\sqrt{3}$ = 1.732
E_L = LINE-TO-LINE VOLTAGE (VOLTS)
I_L = LINE CURRENT (AMPERES)
PF = POWER FACTOR (EXPRESSED AS DECIMAL)

NOTE: Omitting the power factor in the power formula gives volt-amperes:

$$\text{VOLT-AMPERES} = \sqrt{3}\, E_L I_L$$

EXAMPLE:

How many kilowatts and volt-amperes are drawn from a 3-phase source when it is delivering 40 amps at 240 volts to a motor with 86% PF?

$P = \sqrt{3}\, E_L I_L\, PF = 1.732 \times 240 \times 40 \times .86$
$= 14,299$ WATTS $= 14.3$ KW

VOLT-AMPERES $= \sqrt{3}\, E_L I_L = 1.732 \times 240 \times 40$
$= 16,627$ VOLT-AMPERES
$= 16.63$ KILOVOLT-AMPERES (KVA)

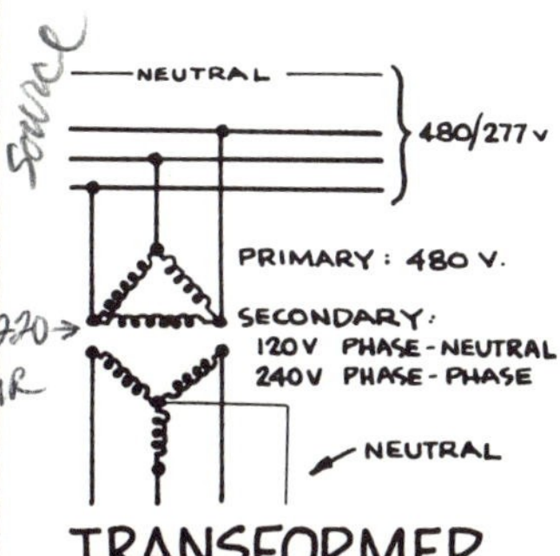

TRANSFORMER *with 4-wire secondary needs only a 3-wire connection to primary from 4-wire source. Source neutral is not required by transformer.*

across the delta-connected transformer primary windings. Then, with a 4-to-1 step-down ratio from each primary winding to its secondary winding, the 120-volt secondary windings are connected in a wye to obtain 120 volts phase-to-neutral and 208 volts phase-to-phase. *It is not necessary* to use a primary wye connection for a transformer fed from a 4-wire wye system.

Three-Phase Power

The power delivered by each phase winding of a wye source operating at unity power factor is equal to Phase Voltage times Phase Current, i.e., $P_P = E_P \times I_P$. The total power delivered by all three windings is $3 \times E_P \times I_P$. But, since phase current is line current in a wye and since phase voltage is equal to line voltage divided by $\sqrt{3}$, the power rating of a wye source at unity power factor can be expressed as—

$P = \sqrt{3} \times E_L \times I_L$ where,
P = watts, E_L = line or phase-to-phase voltage, and
I_L = line current.

The foregoing assumes that all phase windings of the source have the same current rating and the same voltage rating. The same formula is used to obtain the power taken by a three-phase load of balanced current and voltage ratings.

If the power factor of the three-phase load is other than unity, power factor being the cosine of the angle between phase current and phase voltage, the formula for power in watts becomes $P = \sqrt{3} \times E_L \times I_L \times$ Power Factor. For other than unity power factor, the $\sqrt{3} \times E_L \times I_L$ gives the volt-amperes of the system and dividing that figure by 1000 gives the kva (or kilovoltamperes) of the system.

In a delta-connected system, the same formulas are used to obtain power and kva ratings of sources and loads. Basically, the power of a balanced delta system is equal to $3 \times E_P \times I_P \times$ Power Factor. But since phase voltage is equal to line voltage and phase current is equal to line current divided by $\sqrt{3}$, the formula can be expressed as $P = 3 \times E_L \times \frac{I_L}{\sqrt{3}} \times PF = \sqrt{3} \times E_L \times I_L \times PF$.

Y

$I_P = I_L, \ E_P = \frac{E_L}{\sqrt{3}}$

Δ

$E_P = E_L \quad I_P = \frac{I_L}{\sqrt{3}}$

$P_{TOTAL} = 3 E_P \times I_P$ ← phase

$= 3 \frac{E_L}{\sqrt{3}} \times I_L$

$= (\sqrt{3})(3)\left(\frac{E_L}{3}\right) I_L$

$P_{TOTAL} = \sqrt{3} E_L I_L$ ← LINE

Practical Circuits

APPLICATIONS of electrical equipment in modern systems for power, light and heat demand a variety of practical circuit calculations. Typical circuit layout tasks are as follows:

Circuit Loading

Assume a certain large convention hall is to be lighted with 500-watt 120-volt incandescent luminaires recessed in the ceiling. Lighting design indicates that 16 units will provide the lumen output to obtain a required footcandle level in the auditorium. This is a total of 16 x 500 or 8000 watts. Because the ceiling of the hall is very high, it is decided to serve these lighting units from a separate panelboard placed in a room above the ceiling of the hall. With such an arrangement, the lengths of the branch circuits can be kept to a minimum, with a 120/240-volt, 3-wire, single-phase feeder run to the panel.

To minimize the number of circuits and to reduce the size of panelboard required, 30-amp circuits will be used to supply the total load. The total wattage capacity of each such circuit is 30 amps x 120 volts or 3600 watts. It might appear that 7 of the 500-watt units could be used on such a circuit since 3600÷500 is 7 plus. But the National Electrical Code requires that circuits which operate continuously for long periods of time must not be loaded in excess of 80% of the circuit rating. This means the load on a 30-amp, 120-volt circuit must not be greater than 0.8 x 3600 or 2880 watts. To conform to this load limit, only 5 of the 500-watt units can be used on each 30-amp circuit. But voltage drop also has to be considered, so assume that only four luminaires can be used if we want to keep voltage drop below 1%.

If four units are to be used on each 30-amp circuit, the total of 16 units will require the use of four 30-amp circuits. Now a 4-circuit panel could be used, but good design dictates the use of a slightly larger panel to obtain a couple of spare circuits for future load expansion. A 6-circuit panel was chosen to provide two spares.

As shown in the diagram, a magnetic contactor is used in the panel to provide control of the mains. The control circuit of the contactor is then carried down to a pushbutton station on the main floor level of the convention hall, to provide convenient control of the ceiling lighting.

Feeder Sizing

In the above example, a total of 8000 watts of lighting load is being served by a 120/240-volt panel. Since the luminaires are rated at 120 volts, half of them will be connected from one panel bus to the neutral, and half will be connected from the other panel bus to the neutral. There

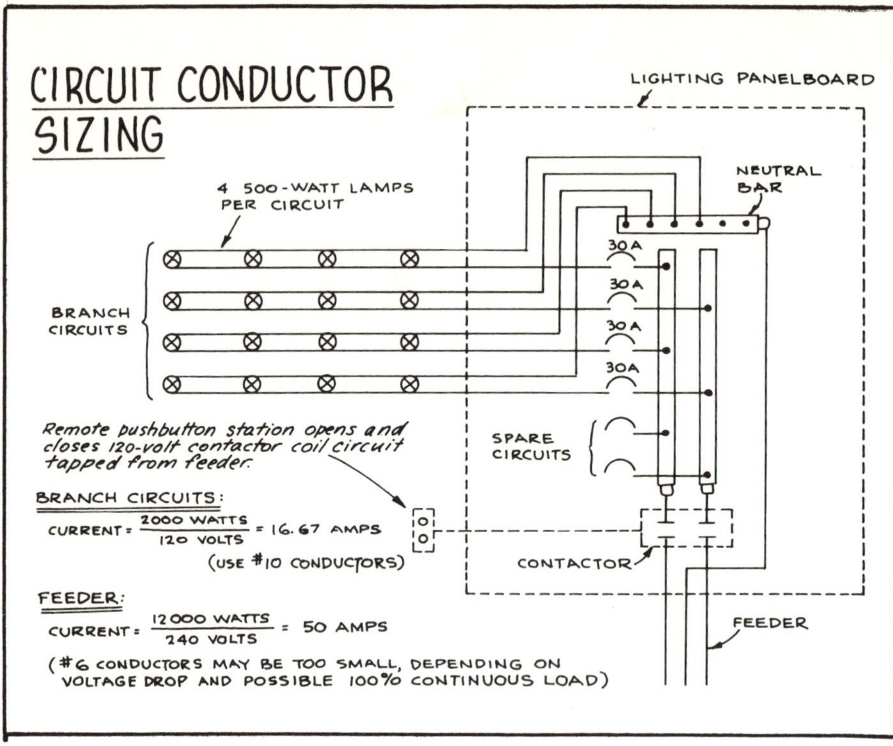

will be two branch circuits connected to one bus and two to the other. This will place 4000 watts on each panel bus, and the feeder conductor to each panel bus must have sufficient current-carrying capacity for such a load. But this would be sizing the feeder conductors only for the present load, with nothing for load growth.

There are several ways in which the size of the required feeder conductors can be determined. Considering each bus in the panel, as in the previous paragraph, the current capacity of the feeder conductor should include the present load, 4000 watts, and should include capacity for addition of future load up to the ability of the present circuits and the spare circuits to handle more load. Of course, inclusion of spare capacity is an arbitrary thing and depends upon the particular conditions in any specific case. Here, assume an additional future load of 2000 watts for the one spare circuit on each bus. This then gives a total of 6000 watts on each busbar, requiring a total capacity of 12,000 watts to be served by the 120/240-volt feeder to the panel.

The required current rating of the phase conductors of the feeder is obtained by finding out how much current must be supplied at 240 volts to equal 12,000 watts. That is, I=12,000 ÷ 240 or 50 amps. This can also be arrived at by considering that the load current from either busbar to the neutral is equal to 6000 watts 120 volts, which is again 50 amps. The neutral conductor of the 3-wire 120/-240-volt feeder must also have a rating of 50 amperes because that would be the load imposed on the neutral if the two circuits connected to one busbar were operating and the two circuits connected to the other busbar were not operating. This latter condition is referred to as maximum unbalance of the load. The NE Code, in Sec.

220-4 (d), says that the neutral must be capable of carrying the maximum unbalance of the load. Under the condition that the 4000-watt load and the future 2000-watt load on each busbar are operating, the two phase conductors (i.e., the two wires with 240 volts between them) will be carrying 50 amps; but the neutral conductor will have zero current flow.

The feeder conductors can then be selected from the Code Tables (310-12 through 310-15) to find the required size of conductor, with specific insulation, rated for 50 amps. For instance, No. 6 types TW or RHW, or No. 8 type RHH might be used for this job. But the Code table only indicates that the wire is suitable for use from the standpoint of temperature rise. Voltage drop would have to be taken into consideration and that will frequently demand the use of larger size of the copper cross section to minimize the voltage drop in the wire due to current flow x resistance of the wire (the IxR drop). The larger the size of wire, the lower is the R and, consequently, the lower is the voltage drop for the given current.

Transformer Calculations

Assume a 75-kva transformer is to be used to supply 120/208-volt loads from a 480-volt, three-phase distribution system. The transformer was selected, say, as the next larger size to a demand load of 60 kva at 120/-208 volts. Sizing of the primary and secondary feeder conductors is done according to the formula kva $= \sqrt{3}$ x E x I/1000. From this, I = kva x 1000/$\sqrt{3}$ x E. Primary current rating of transformer—I = 75kva x 1000/-1.73 x 440 = 99 amps. This would require conductors not smaller than No. 1 TW or No. 3 RHW. Secondary current rating—I = 75kva x 1000/$\sqrt{3}$ x 200 = 214 amps. This would require conductors not smaller than 250 MCM TW or No. 4/0 RHW.

Note that low values of line-to-line voltage were used (440 and 200) to obtain the higher current values for sizing feeder conductors.

TRANSFORMER CONDUCTOR SIZING

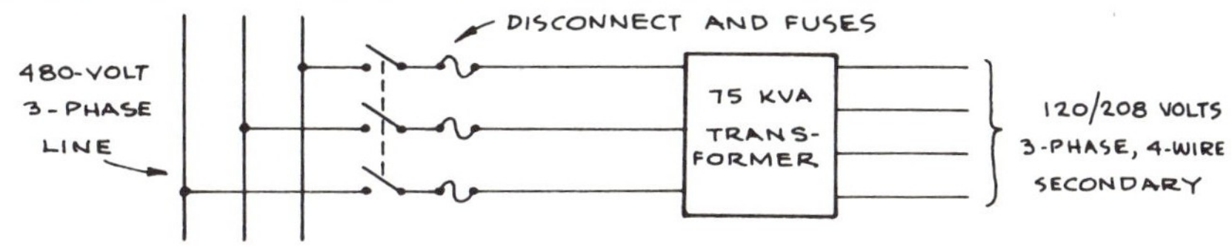

3-PHASE VOLT-AMPERES = $\sqrt{3}$ x VOLTAGE x CURRENT

$$\text{3-PHASE KVA} = \frac{\sqrt{3} \times \text{VOLTAGE} \times \text{CURRENT}}{1000}$$

$$\text{THUS CURRENT} = \frac{\text{KVA} \times 1000}{\sqrt{3} \times \text{VOLTAGE}}$$

For current calculations, voltage values are deliberately chosen lower than rated to assure adequate conductor capacity under possible low system voltage conditions.

PRIMARY:

$$\text{CURRENT} = \frac{75\text{ KVA} \times 1000}{1.73 \times 440\text{ VOLTS}} = 99\text{ AMPS}$$

(USE #1 TW OR #3 RHW)

SECONDARY:

$$\text{CURRENT} = \frac{75\text{ KVA} \times 1000}{1.73 \times 200\text{ VOLTS}} = 214\text{ AMPS}$$

(USE 250 MCM OR #4/0 RHW)

INDEX

1 HP = 746 WATTS

= 550 FT-LBS / SEC

≈ 3/4 KILOWATT

1KW = 3,413 BTU/HR

LIQUID → ICE   144 BTU/LB (LATENT HT OF FUSION)

LIQUID → STEAM   970 BTU/LB (LATENT HT OF VAPORIZATION)

1 TON OF REFRIGERATION = 12,000 BTU

400 cfm / ton

$cfm = Av$

$BTU/HR = UA\,\Delta t$